The Boys
in the Cave

If you'd like to keep track of the
titles you've read,
please don't mark up our books
– mark this paper instead.

The Boys in the Cave

Deep Inside the Impossible Rescue in Thailand

Matt Gutman

An Imprint of HarperCollins*Publishers*

Original survey of Tham Luang (p. ix) by Association Pyrénéenne de Spéléologie (1987). Surveys of Monk's Series, Tham Nang Non series, and Main Cave extensions by Vern Unsworth, Rob Harper, and Phil Collett (2014–2016). Composite survey of cave system by Martin Ellis (2018).

HarperCollins books may be purchased for educational, business, or sales promotional use. For information please e-mail the Special Markets Department at SPsales@harpercollins.com.

FIRST HARPERLUXE EDITION

ISBN: 978-0-06-291071-4

HarperLuxe™ is a trademark of HarperCollins Publishers.

Library of Congress Cataloging-in-Publication Data is available upon request.

18 19 20 21 22   ID/LSC   10 9 8 7 6 5 4 3 2

For Saman Gunan, who sacrificed everything
and
Paul Gutman, for whom rescue was impossible

Contents

Part Two

THAM LUANG CAVE COMPLEX

Tham Luang Forest Park, Pong Pha, Mae Sai, Chiang Rai, Thailand | 47Q 590619 2253976 Alt.: 446 m | Length: 10,316 m VR: 85 m

ALL CHAMBERS

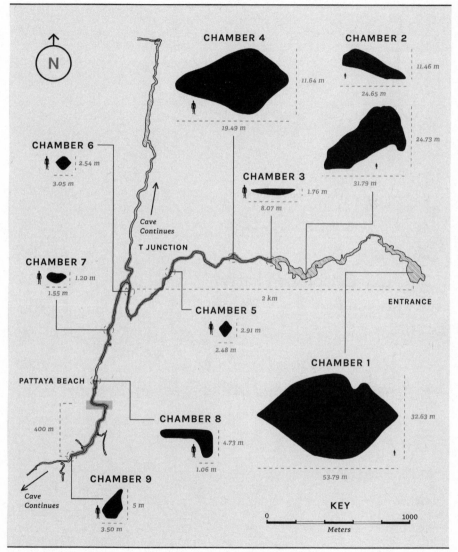

N

CHAMBER 4
11.64 m
19.49 m

CHAMBER 2
11.46 m
24.65 m
24.73 m
31.79 m

CHAMBER 6
2.54 m
3.05 m

CHAMBER 3
1.76 m
8.07 m

Cave Continues

T JUNCTION

CHAMBER 7
1.20 m
1.55 m

2 km

ENTRANCE

CHAMBER 5
2.91 m
2.48 m

CHAMBER 1
32.63 m
53.79 m

PATTAYA BEACH

400 m

CHAMBER 8
4.73 m
1.06 m

Cave Continues

CHAMBER 9
5 m
3.50 m

KEY

0   Meters   1000

MAP INFOGRAPHIC CREATED BY: CHIANG MAI ROCK CLIMBING ADVENTURES, LTD., 2018   www.thailandclimbing.com
Surveyed by: Association Pyrénéenne de Spéléologie (1986-1987) Grade UISv2 3-3-A/2-2-A
Map drawn by Martin Ellis (June 2018)

The Boys
in the Cave

Prologue

They had surfaced in alien territory. A three-hour hard swim from help, they were suspended in a tunnel darker than the remotest corner of space, where no radio or cell signal could penetrate. It was a spot so distant and hostile it had repelled rescuers for nine straight days—it was in northern Thailand, but it might as well have been on another planet.

For hours, British cave divers Rick Stanton, fifty-seven, and John Vollanthen, forty-eight, had finned against the current, breathing heavily in their scuba regulators and carefully unspooling the pencil-thin guideline behind them. To keep the blue line more or less in the center of the passage, they would tie it around a hanging stalactite or a dagger of limestone jutting into the tunnel. It had continued like this until

Stanton glanced down at the 250-yard spool, noticing that it only had about five yards left—which meant they had a decision to make. That guideline was basic diving protocol, and it was their sole link to the outside world—ensuring they wouldn't get lost or trapped on the mile-and-a-half-long journey out of the cave.

The cave-diving pair was now farther into the Tham Luang cave than any of the hundreds of rescuers before them had been. To survive they could rely only on what they carried: the three lights attached to their helmets, a few zip ties, their fins, a chocolate bar in a wet-suit pocket, and the two air cylinders each. Stanton checked the dial of his air gauge; he had consumed about a third of his air supply, another reason to turn back.

If cave diving was a religion, the rule of thirds would be its first commandment: use a third of a tank on the journey in, save a third for the journey out, and always reserve a third in case of trouble—like getting lost even in well-traveled passages, becoming physically stuck in chokepoints where the cave pinches down from the size of a dining room to a dinner plate, getting tangled in their own guideline; all of these can lead to a diver running out of air and dying. Quicker deaths can result from equipment failure, landslides, flash floods, or just slamming headfirst into rock. And sometimes the end is generated from within—the boundless darkness itself leading to panic attacks, and panicking in the water often leads to death.

It was Monday, July 2, and the two divers—skinny, pale, middle-aged Brits—were not planning on dying that day. They were among the world's best cave divers, comprising the two-man tip of a spear in a rescue operation involving up to 10,000 soldiers, civilian rescuers, and volunteers. Twelve members of a local Thai youth soccer team and their twenty-four-year-old coach had gone into the cave on June 23 and had never come out. The cave opens with a series of hangar-size rooms and tapers down into passages as tight as the space between your car and the ground. In their search, rescuers had found ample evidence that the boys were in the cave: the backpacks they had dropped, cleats, their bikes parked outside. But they had zero evidence they were alive.

Three hours earlier Stanton and Vollanthen had set out from what was known as Chamber Three, the dive launch point inside the Tham Luang cave in northern Thailand. The region, wedged between Myanmar and Laos, is known as the "Golden Triangle"—a term coined by the CIA, not for the region's lazy sunsets, but for its centrality as a hub for the opium trade. Thailand wiped out the opium trade on its side of the border in the late 1950s, but the exotic name for the region, with its mist-crowned mountains, numerous tribal villages, and dense jungles, stuck. A fellow Brit, Vernon Unsworth had been drawn to the region by other attractions: its many caves. He knew Tham Luang better than anyone in the world and had hand-sketched a map for Stanton and Vollanthen. They had committed it to memory. It was believed that the boys might have retreated beyond a part of the cave locals called Pattaya Beach—after the famous Thai resort town. It had a high sandbar that typically remained above water even during the floods. As the divers swam past it that day unspooling their rope, they noticed that the floodwaters pulsing through the cave had swallowed up Pattaya Beach. Unsworth's guess was that just a few hundred yards deeper into the cave beyond Pattaya Beach there was a side room that offered high ground, and that the boys might be somewhere around there. At least that's where he'd go.

People who so habitually expect trouble that they keep a third of their most precious commodity in reserve are naturally pessimistic. Cavers who don't anticipate the worst often don't survive long. Stanton and Vollanthen were veterans of multiple cave rescues in which they had sometimes brought people out alive;

more often than not, they found corpses. To their knowledge no one with zero provisions had survived this deep in a cave for this long. They figured that, sadly, wherever these boys were, they weren't alive.

Whenever the divers noticed the cave ceiling rise enough to reveal an air pocket above, they would inflate their buoyancy vests and kick up to the surface. That they didn't have maps with them and couldn't get a GPS reading or communicate with the outside world mattered little to them. Decades of experience told the pair their most important tool was already attached to their faces. For the past several hundred yards, each time they noticed those air spaces they'd bob up and take a sniff—their noses supplying information their eyes couldn't. Each time, Stanton would remove his mask, and take a couple of quick nasal inhalations—sample the air. Before he took his mask off this time, Stanton made a mental note to tell Vollanthen they should turn around soon because they were running out of air and guideline. And then he sniffed.

This may have been terra incognita, but the smell—he recognized that immediately.

It's the distinct smell of human shit, he thought at first. Then, continued his short internal monologue, *it's so* very *pungent, so overwhelming, it might actually be the smell of decaying bodies.* He nudged Vollanthen. "Hey, John. We've got them. Or got something. Take your mask off and confirm."

As the Brits began debating whether the noxious odor was the product of excrement or corpses, they heard voices. As they drifted toward the smell and the sound, a beam of light flicked on and scanned the water.

Moments earlier, their twenty-four-year-old coach, Eakapol Jantawong, had heard something: men's voices. They all had. The boys who had been digging stopped cold and the coach asked everyone to hush up. Silence. Then the voices again. The coach told twelve-year-old Mick, who was holding their flashlight, to go down to the water's edge to check it out. But the boy froze with fear and didn't move. The coach whispered, "Hurry, go quickly. If it's a rescuer they might pass us."

The boys were unsure if what they were hearing "was real." They had gone ten days without food, and so zealously husbanded their flashlight batteries that they spent most of the time in complete darkness. They were over a mile and a half into the Tham Luang cave; directly above were six hundred yards of limestone rock. Not a single photon of light penetrates this place—so when flashlights are switched off, there was nothing for the rods and cones of their retinas to adjust to. The darkness was complete. And lately, the boys crowded into their bathroom-size living area above the waterlogged passage had been straining to listen to a chorus of inexplicable sounds—dogs barking, roosters, even children playing. Hearing these new sounds, fourteen-year-old Adul Samon snatched the flashlight from Mick and moved toward the water.

The boys saw lights, and two creatures that looked like spacemen with strange breathing hoses seemingly ripped from a car engine attached to their mouths and helmets bristling with lights. The semi-submerged figures were talking and, cautiously, the boys slid down the slope to greet them.

"Officer! Officer, hello! Over here!" they called out in Thai. The voices didn't answer.

Adul, already stupefied that they had finally been found, was doubly confused when he realized the men were not speaking Thai. It was . . . English. He crept to the water's edge. His mind sluggish after more than two hundred hours without food, all he could muster was a warbly "Hello!"

The divers had surfaced about fifty yards away from the boys. By twenty or so yards out, their headlights illuminated a couple of the boys. They were relieved—at least *some* were alive.

PART ONE

Chapter One
The Moo Pa

It had started out as a pretty typical Saturday in Thailand's northernmost town, Mae Sai, snug against the Myanmar border. At about ninety degrees, the air was a hot damp towel wrapped around them, but the boys practiced anyway—they always did.

Most of the boys cycled to the pitch. They turned off the country's Highway Number 1 and rode the gentle incline toward the spine of mountains separating Thailand from Myanmar. Then past the market on the southern side of the road, with vendors hawking stinking anchovies and rows of bok choy, cilantro, and ripening pineapple, for which the region is known. Up past the Wat Ban Chong temple, with its gilded Buddha crowned with neon lights. One final leg took them up the unnamed road past the honky-tonks that doubled as brothels, ringing out with midday karaoke and the croaks of beer-fattened drunks.

The pitch was on a rise, above all of that. Shaved flat from the top of the hill, it had seen better days, back when you could easily make out the boundary lines. The heat and the rains had nibbled at the now ratty goal nets and bitten off chunks of the concrete viewing porch.

They'd spent a lot of time there, the Wild Boars. The team's head coach, Naparat "Nok" Guntawong, founded the team in late 2013, and they'd done Mae Sai proud ever since. A former semiprofessional midfielder, Coach Nok remained boyish and slim—even when he wasn't anywhere near a playing field he invariably wore the polyester mesh of athletic clothing. The telltale sign of his profession as a shipping agent were his yellowing, half-inch-long pinky nails. Coach Nok had hoped to infect his daughter with the soccer bug, so a few times a week he would take her to kick the ball around. Sometimes her male school friends—who were much more interested in the "beautiful game"—would join them.

In late 2013, those same boys got together ahead of a local youth tournament and asked Nok to coach the fledgling club they wanted to form. The name Wild Boars evokes sylvan ferocity. It conjures images of muscled young men, recklessly gunning for the goal. Sure, the team was ferocious, but it was actually named after its sponsor, which exports live pigs to China. In keeping with the company's porcine theme he figured: we're near the forests, boars are tough, why not *Moo Pa*, which translates to Wild Boars? It was close enough for the sponsor.

The Wild Boars, eighty-four strong, were separated into three age groups: under-thirteens, under-fifteens, and under-seventeens. They were short on finances but long on grit and teamwork: They'd practice up to twenty hours a week, hitting the pitch several days a week for a couple of hours and then on weekends often spending whole days together. Drilling and scrimmaging would last most of Saturday morning, and the rest of the day would often be spent on team-building exercises. Since many of the boys had bikes, sometimes they'd ride places with Coach Ek.

If Head Coach Nok was the general, Coach Ek was his friendly lieutenant—with his smiling eyes and chirpy voice, he was more big brother than drill sergeant. He had the perfect temperament to train the under-thirteen group, which he had been doing for about three years. The veterans of the team would sleep over at the modest home he shared with his aunt. He was very close with some of the boys; often when eleven-year-old Titan's parents went out of town, they'd leave Titan in the care of the coach.

Before big games, the former monk would lead them in the Buddhist meditation practice of vipassana, which focuses on mindfulness breathing and the understanding of the ever-changing and impermanent nature of reality. When he wasn't coaching he would do odd jobs at the Wat Doi Wao temple, where he had been a monk. The temple practiced the central Buddhist pillar of "loving-kindness," or benevolence to others. The monks were gentle with all of nature's creatures—especially humans. Harsh words were rarely spoken to the boys, and corporal punishment was taboo. In many ways, Coach Ek was the focus of many of the boys' social lives, and he would be seen tousling their hair and joking around with them.

Perhaps the most striking thing about Coach Ek was his cheeriness despite an early life marked by suffering. A member of the Tai Lue minority that roams the mountains between Thailand, Myanmar, and Laos, he was born stateless and remained stateless. Thai was not even his first language. His family had been part of the working poor, his father a cook in a local restaurant; when an epidemic swept through their village sometime in 2003, it first claimed his little brother, then his mother, then his father.

A couple of years after his parents died, the coach's aunt sent him to a monastery. For centuries, Buddhist temples in Southeast Asia have attracted poor young boys seeking free education and steady meals. They are taught their letters, but also Buddhist discipline. For the next decade, as he learned the teachings of Buddhism, Ek's stomach had growled between the noontime meal of rice and a soup or stew and the next meal the following morning. Buddhist monks in Thailand believe in eating only twice a day, a meal at sunrise and another at noon. Anything more would be decadent. This daily fasting isn't predicated on self-flagellation. Rather, Buddhist monks believe that deprivation enables them to focus on their practice, on meditation. The goal, handed down from the teachings of the Buddha himself five hundred years before Christ, is to unshackle monks from the constraints of desire. The Buddha knew how we all feel before dinner, craving that burger or nightly tot of whiskey. Strip that away and you can focus. So after the low rumble of pre-meal chants Ek—like his fellow apprentice monks sitting on the floor in front of wide knee-high tables— would ladle fat mounds of steaming rice into his bowl.

After nearly a decade as a monk, he now enjoyed the freedom of life beyond the temple. One of his special rites of passage was taking the boys to the Tham

Luang cave, about a half-hour bike ride from the pitch. A week earlier, some of the team members who had formed an intra-team bicycle group posted messages on their Facebook page that they'd be riding out to the cave on Saturday. The cave offered a refuge from the simmering heat and—especially appealing to Ek, the former Buddhist monk—detox from the IV drip of cell phone signals upon which the boys were hooked. The cave walls jammed those jangly ringtones and the chirps of incoming messages until only silence and togetherness remained.

On Saturday, June 23, Head Coach Nok had livestreamed part of the team's intramural scrimmage on Facebook, and then had gone home, unaware that some of the boys had planned to go to the cave with Assistant Coach Ek. Once practice had ended around noon, the boys ducked under the rusted rail surrounding the pitch and crossed the street to a tiny unlit shop, just as they did every weekend. The smiling old lady who runs the shop was there when they arrived, selling them Lay's potato chips, Dino Park (dinosaur-shaped fritters), and a savory snack called Bento—leathery strips of desiccated squid. They drank Pepsi and a popular yellow-tinted electrolyte drink called Sponsor and prepared to head out to the cave.

A few of the boys had been there before, but it

was birthday boy Peerapat Sompiangjai's first time. Many Thais choose their own, shorter nicknames; Peerapat called himself Night. He was excited for the adventure—the only hitch: they'd have to cut the excursion short, because Night was due back home around 5 P.M. for a birthday celebration. The cake, decorated with a big toothy emoji, was already cooling in the fridge.

As the journey to the cave got under way, Coach Ek, in a bit of precarious, don't-try-this-at-home-kids multitasking, live-streamed the boys' bike ride as he rode on the back of a moped driven by the team's fourteen-year-old goalie Biw. In the video, the moped zooms by the boys in a flash of red and blue jerseys before the camera pans back to show them all pumping the pedals in their flip-flops or sneakers, their nylon backpacks stuffed with clothes, soccer cleats, and shin pads. The road narrows from there, and, cackling with delight, they are not at all winded by the additional trek after a long practice. These are, after all, the Wild Boars. Coach Ek's camera bounces and the image pixelates as the road roughens. The battleship-gray barracks of a local military base give way to creaky two-story buildings anchored by first-floor shops, with apartments above. Buildings yield to jungle as the blacktop switches to dirt track. A pack of raggedy dogs howls a warning. They're nearing the cave. Banana trees, cane, and tamarind—with its thorny branches protecting seedlings—hug the roadside. Green jolts of grass spring from every crack in the earth. Egrets form dots of white against the green wall of the mountain. And as they head uphill toward the cave and the phone camera tilts upward, it catches a few fleeting frames of the dark clouds crowding Doi Nang Non mountain, which translates roughly into "the Sleeping Princess."

If you're driving from south to north on Thailand's Route 1, you can't miss the ridge of mountains that divides the country from Myanmar. Five miles south of the town of Mae Sai, look up and you might notice, if you squint, the distinct profile of a supine woman. From north to south, there's the sweep of hair flowing down toward a dramatic escarpment. Direct your eyes slightly southward, to your left, and you'll make out a prominent forehead, the gully of her eye socket, and the rise of a prim little nose. Farther south, her chin dips into her neck, then her torso rises to an ample bosom. South of that, where her pregnant belly would be, the mountain soars seemingly twice as high as the other ridges. For those who have seen it, it's like an autostereogram—once your eyes have solved the puzzle, you can't stop seeing it.

Tham Luang Nang Non roughly translates to "the

great cave of the sleeping princess." The legend goes that a beautiful princess fell in love with a stablehand in her father's kingdom. She became pregnant. The tryst was forbidden and the king became enraged. The young couple fled, seeking shelter in the not-exactly-homey cave. When her lover was out foraging for food, the king's men found and executed him. In her grief and rage, the pregnant princess stabbed herself with a dagger. Local folklore says that her body morphed into the mountain and that the stream flowing through the center of the cave is her blood.

Spawned by the mix of Buddhism and ancient local animism, the tale is similar to many others around the world, where people have devised legends to explain unusual or possibly dangerous phenomena; in this case a human-shaped mountain crest and a colossal cave beneath are endowed with host spirits to be revered and feared. As in the great caves of Mexico's Oaxaca region, taboos are established around the legends—along with a cast of easily enraged spirits who frequently need to be appeased with offerings or sacrifices. But it didn't take an ancient shaman to understand that the cave with the giant mouth and increasingly narrow guts was dangerous, especially in the rainy season.

When the boys arrived, they rested their bikes in the bushes near the cave and dropped their gear. Some kicked off their shoes and left any unnecessary encumbrances behind. They didn't have to worry about carrying the heavy soda cans and glass bottles or their snacks of squid and fried batter; they'd scarfed them down before they left the pitch. One boy left his training pants dangling over handlebars, as if hanging to dry. Theft isn't a problem in this part of Thailand, and anyway few people come to the cave during this time of year. Besides, the boys wouldn't be gone long. They walked down a set of muddy stairs past a spirit temple housing a trio of mannequins in pink silk, then up another flight of steps past a second altar housing a plaster idol of the Sleeping Princess herself. Near the mouth of the cave was a sign that read, in Thai and English, DANGER!! FROM JULY TO NOVEMBER THE CAVE IS FLOODING SEASON.

It was only June 23; there was nothing to worry about.

Chapter Two
Some Birthday

They had only planned to go in for a short time. After all, as everyone knew by now, Night had to get to his birthday party. The emoji cake was waiting. His grandmother, parents, and kid sister were getting the grill ready for his favorite meal: grilled pork and shrimp. It was only his first season on the team—he'd only been with the guys for two months—but he'd gone with them because he wanted to take part in the adventure. They had their flashlights, and some brought in their backpacks—but they wouldn't need much for such a short sprint in and out of the cave.

They climbed those stairs, past the warning sign about flooding season beginning in July.

They proceeded in. Coach Ek led the way, the backpack slung over his shoulder containing a length of green rope and extra flashlights with batteries. Behind him were Night, as well as Tee, a fifteen-year-old captain; fifteen-year-olds Note and Nick, who was a bit of a ham and Night's cousin; fourteen-year-old Biw (pronounced "Beeyoo"), the goalkeeper with the moped, round-faced and tall for his age; fourteen-year-old Adul, who, with a hint of dark peach fuzz over his top lip and a more muscled physique, looked older than many of the other boys; Tern, also fourteen; thirteen-year-olds Dom (another captain), Pong, and Mark; and twelve-year-old Mick. Giggling among them was the little guy, ironically nicknamed Titan, who was eleven.

Given the enormous differences in physical development and socioeconomic backgrounds—Adul, Tee, and Night pretty much looked as old as the coach, while Titan was a grinning pipsqueak—they were a remarkably cohesive group. Mark, Adul, Tee, and Coach Ek were stateless. Adul, a refugee from nearby Myanmar, had been living at the Mae Sai Grace Church for a decade, and was the recipient of a scholarship to a local private school. None of the boys seemed to care much about economic status, or religion for that matter. All but Adul were Buddhist. Soccer and a sense of adventure united them. The big boys took care of the little boys—including Mark, who was thirteen but the same size as Titan—and the little boys tried to keep up. The coach, who was shorter than several of the boys yet powerfully built with ropey arms and tree trunks for legs, took care of them all. Heading in they luxuriated in the blast of chilly air.

They ascended two flights to the cave's oblong opening—which looks like a gaping mouth baring five-foot-long mossy teeth. The mouth exhales gusts of cool, musty air revealing the cave's grand lobby, which would befit the grandest of five-star hotels; it is, in fact, big enough to fit the Taj Mahal. The boys who had never been there before struggled to comprehend the enormity of Mother Nature's creation. In that first chamber, chandeliers of stalactites somehow affixed to the cave ceiling dangle from above. Below and to their right was a large, dusty gravel bed and mud stains twenty feet up showing the high-water mark of the previous year's flood.

The path through the first chamber, which the cave's recent British explorers had named Mae Sai High Street, is well traveled and clearly marked with a rusty metal hand rail. Farther in it snakes around obstacles and rougher terrain. Even farther in, as the colossal first room slims down, the cave lobby is draped with stalactite curtains dozens of feet high—formed when stalactites, which hang from the ceiling, meet stalagmites, which are basically drip castles formed by drops falling from stalactites above. Hundreds of millennias' worth of droplets laced with minerals have deposited microscopic calcium rings that year after year build out the stalagmites in ultra-slow-motion. The color scheme seems as if it's pulled from a 1970s motel: brick-brown, off-white, and fungus grey—a pattern of prints fashioned by the various minerals absorbed by the water as it filters down fifteen hundred feet from the tree canopy, through the soil, and into the porous limestone. Because some of the boys had been there before, they knew where they were going: a chamber in the cave nearly three miles in called Voute Basse, which means "low vault" in French. Though it's not particularly impressive, Voute Basse, or as the kids knew it "the Underwater City," was the cave's terminus for most amateurs. Just beyond, covered by a low overhang, is a deep pool of water that is flooded even during the dry season—going farther required scuba gear, or squeezing through a tiny "window" to the skinny passage beyond that only expert cavers knew about. It is there, on the walls just before that pool, where Mae Sai's most intrepid boys, often on the occasion of their high school graduations, would scribble their names.

The boys, some barefoot, padded on, heading nearly due west, bound for Voute Basse. If the tunnel had continued in that direction for a little over a mile, they would have found themselves beneath the border with Myanmar. About two hundred yards in they encountered the first squeeze, as the grand lobby tapered off to a passage roughly the size of the crawl space you might find under a staircase. The boys stooped to get through and kept going, marching fast. Little Titan, who was experiencing the cave for the first time, was scared. He had been with the team for two years, and had begged his mother to let him join, but as one of the youngest he found himself afraid of the dark and the creepy shadows cast by their flashlights, He didn't dare tell the older boys though; they were covering ground quickly, not stopping to take in the views or rest much.

They encountered more chambers the deeper they went in. The grand, nearly seventy-five-foot-high Chamber Two is also impressive. Thirteen flashlights raked the rocky dragon's teeth above, producing a discotheque's strobe effect. Beyond that the tapering continued. The passage to get into Chamber Three was just high enough for them to stand in, if they walked on the lowest, gravelly part. They had to crab-walk for the next 150 yards or so until they hit Chamber Four, with thirty-foot ceilings. It's the last of the grand rooms. Knowing the route ahead might get a little wet, the boys who had carried their backpacks inside dropped them here, while others kicked off their knock-off Adidas slides, continuing barefoot. The cave floor there was dry, cracked clay that hadn't seen significant moisture since the last monsoon season.

The boys now were forced to walk and crawl mostly single file through the tunnels. They mushed to a T-junction intersection about two thousand yards into the cave. Turning right leads north and slightly uphill to a cave extension called the Monk's Series—named by the French cavers who first mapped Tham Luang for a little meditation structure that has long since vanished after many flood seasons; turning left, or nearly due south, leads to Voute Basse—the Underwater City.

A few thousand feet up from that right-hand turn toward the Monk's Series is a little stream that drains down into this extension, slaloming between bamboo stands and fishtail palms. While the exact draining point is a mystery, the amount of water it feeds into that narrow channel is not. When it flows, picking up momentum as it heads downhill toward the T-junction, it has the volume of a decent-size river.

But the boys and their coach couldn't have known that, nor could they have known that outside the cave the rain had started. And they certainly were not aware that since this year had already produced a foot of water more than usual, the mountain wasn't as thirsty as it

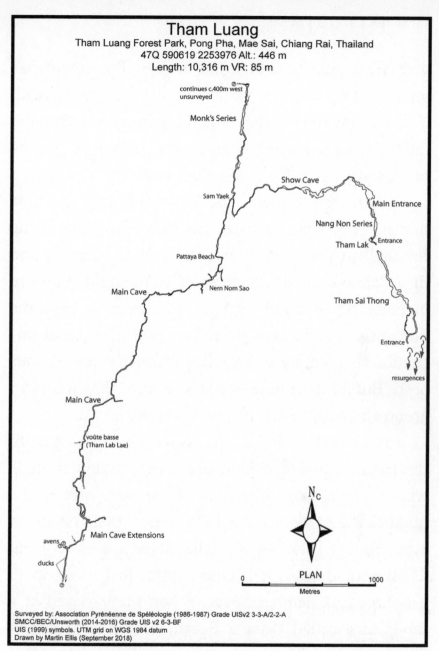

Tham Luang

Tham Luang Forest Park, Pong Pha, Mae Sai, Chiang Rai, Thailand
47Q 590619 2253976 Alt.: 446 m
Length: 10,316 m VR: 85 m

continues c.400m west unsurveyed

Monk's Series

Show Cave

Sam Yaek

Main Entrance

Nang Non Series

Tham Lak — Entrance

Pattaya Beach

Nern Nom Sao

Main Cave

Tham Sai Thong

Entrance

resurgences

Main Cave

voûte basse
(Tham Lab Lae)

N$_C$

Main Cave Extensions

avens

ducks

PLAN
0   1000
Metres

Surveyed by: Association Pyrénéenne de Spéléologie (1986-1987) Grade UISv2 3-3-A/2-2-A
SMCC/BEC/Unsworth (2014-2016) Grade UIS v2 6-3-BF
UIS (1999) symbols. UTM grid on WGS 1984 datum
Drawn by Martin Ellis (September 2018)

This is a basic survey of the Tham Luang Cave. It is about 2,000 yards from the entrance of the cave, on the right, to Sam Yek (the T-Junction). Then another 400 yards or so to Pattaya Beach and about 300 yards to Chamber Nine, which does not appear on the map but lies just where the main passage turns south again and near the annotation "Main Cave" at around the middle of the page.

normally would be at that time of year. The mountain's soil would no longer act as a giant dry sponge; instead, it would now repel water, which began pulsing through the hidden cracks in the mountain, pulled by gravity to the lowest possible place. Places like the T-junction.

No, the boys were not aware of any of those things. They were aware of the bats. The glare of the flashlights woke them from their daytime slumber and they flapped around, trying to flee the light. Bats are momentarily blinded by light and seem to have the propensity to zing straight toward you. It's a most uncomfortable feeling when a bat grazes the top of your head. But for most of the boys, as long as they were in a group and with Coach Ek, they weren't afraid.

If you could see a sliced cross section of that part of the passage, it would look like a very rough triangle with a frown-shaped bottom. If the boys kept to the middle, they'd be less likely to whack their heads on a protruding rock between the larger spaces. All but Titan and Mark, the smallest boys, had to stoop to get through Chamber Seven. Then they arrived at a sandy area called Pattaya Beach, named after one of Thailand's best-known resort towns. To the left of this area, a sandy bank stands several feet above the lower parts of the tunnel. A few minutes from Pattaya Beach, which was part of Chamber Eight, the route jogs to the west toward Myanmar, then curves sharply south, where there is the biggest dropdown of the cave—about thirty feet. At its bottom, the boys likely forded what the British explorers called the Goolie Cooler. A goolie is a euphemism for your tender parts—because when you dip in, it's "very uncomfortably cold."

Coach Ek canvassed his boys.

"We might have to swim," he said. "You could get wet and cold."

Tee, the captain, volunteered to go in first. Gingerly he went out into the coffee-dark water.

"Coach, it's not really that deep," he called out. A few moments later, everyone heard him say, "I got to a sandbank, you can all come."

To ford this water, the smaller boys like Mark and Titan hitched a ride on the bigger boys' backs. A short time later they finally reached the Underwater City, where the boys poked around for a few minutes. But there wasn't that much to see. Coach Ek checked his watch—there was no more time for exploring today; it was time to turn back.

The boys were jubilant; for most this was the kind of adventure they had hoped for. Adul, the refugee from Myanmar, who sang for his church choir, played guitar, and craved any new experience, loved it—the otherworldliness of this place, the sense that they were exploring new territory, and the camaraderie. Each boy likely also daydreamed about activities that would follow the little expedition—Night's birthday bash, dinner with the family, maybe a little World Cup viewing. Though they might not have realized it, they had covered the bulk of the cave in excellent time—better than that of some experienced cavers.

But as they marched back home toward the cave's entrance, just a short distance before the T-junction they encountered another body of water. They heard it rushing before they saw it.

"Coach, we found water!" yelled one of the players.

"Are we lost?" shouted another.

Then their flashlight beams bounced off what seemed like a mirage—black water, and a lot of it. It was disorienting. Had they not just come from water about a mile back? A pooling body of water like that was a clear landmark. And it had not been there when they first came through. Coach Ek stopped to think for a minute. He'd been in the cave many times. There was no way they could be lost.

"There's only one path, we aren't lost!" he called to his boys.

Maybe not lost, but certainly stuck. This looked deeper and was moving faster than any of the stagnant water they had crossed so far. So the coach pulled a length of rope from his bag. He tied it around his waist and instructed three of the bigger boys, Night, Adul, and Tee, "If I yank the rope twice, pull me back. If I don't pull it, I made it out and you can come."

He started swimming on the surface and then dove down. He swam by touch rather than sight. Even if he'd had a waterproof light, the water was so laden with silt, he wouldn't have seen much anyway. He felt the larger rocks and the sandy bottom beside them. He wondered if he could dig his way out, unplug a hole. But the darkness and the depth and the tugging current had defeated him. He yanked twice. Night felt panic surge up inside of him as he helped haul in his coach. His heart raced so fast he could feel the rhythmic thumping in his ears.

There was no way out. They could only retreat.

It was now about 5 P.M. Thirteen-year-old Mark started getting nervous. Not about being marooned in this watery wasteland, but because his mother would scold him for being late. Night had to be home for his party, and Titan had exams the next day. Silently, they all knew none of this was happening. The boys' questions started becoming more shrill.

"Coach, how are we going to get out?"

"Another way," he responded.

He suggested they start digging, so the thirteen scrambled to find rocks to use as tools and started carving out a channel to divert the water.

An hour or so later, Tee sidled over to Coach Ek and whispered that they should probably find a place to bed down for the night. They'd been in the cave for hours. After the practice, the ride to the cave, and their nearly five hours inside, the boys were sapped. They hadn't eaten in hours and most everyone was scared. Coach Ek was desperately afraid they would panic, so he told them something he didn't believe himself, announcing that the water was likely a tidal phenomenon and would probably recede by morning.

"You'll see," he promised them. "Why don't we find a place to sleep," he suggested.

They headed back to the softer sand of Pattaya Beach, a couple of hundred yards from the T-junction, as the water continued to pile up. He gathered the boys to pray before they clumped together for sleep. The low chants were his nightly practice, and he was sure it would soothe them as well as himself. It did temporarily distract them, but when the chants stopped it was the sound of the boys' sobs that echoed off the cave walls.

Chapter Three
"How can I sleep when my son is inside?"

Coach Nok was just sitting down to dinner when his phone began lighting up.

"Where are the boys?" a mother asked the Wild Boars' head coach.

"Why isn't Night home?"

"Are you still practicing?"

At this point Nok was mostly irritated, because, as he would insist in the weeks and months following, he had no idea of the boys' plans to go into the cave. Even though the cave is less than two miles from his soccer pitch, Coach Nok had never been beyond its main threshold—a well-traveled chamber which could easily fit a 747. The hole in the mountain is big enough for a little daylight to muscle its way into most of that first chamber. Still, Nok was never tempted to go farther into the narrowing passages beyond that nearly two-hundred-yard-long room—it was too dark, the purchase too slippery, too many spirits to tangle with.

Nok started calling the other boys in the soccer team's bike group, soon connecting with a thirteen-year-old boy named Queue. With a set of jug ears and a short brush of black hair that hung forward slightly over his forehead, Queue was on the Wild Boars and went to school with several of them as well. Their school, called Mae Sai Prasitsart, with a student body of twenty-eight hundred, looks more like a colonial military campus than a high school. The boys wear traditional khaki shorts with high socks and matching blouses topped with maroon epaulets. The school's entrance gives way to a concrete courtyard about the size of a college quad. In front is a raised platform centered around a towering Thai flag, where the school provost in military uniform addresses the kids every morning. The children sit cross-legged on the concrete in perfectly lined rows as he speaks to them and then leads them on a short meditation. Behind its main buildings there is a little pond and garden, which students tend as part of their ecology curriculum. The rows of two-story, lima-bean-colored buildings house classrooms with modern whiteboards, neat rows of newly purchased writing desks, and uniformed teachers.

The students of Prasitsart are mostly good, middle-class kids attending a good school—the future of little towns like Mae Sai. Their families are proud Thais, the kind of people who hang framed portraits of the king and queen in their living rooms along with photos of graduation ceremonies and family trips to the capital city of Bangkok. As is perhaps universal for middle-class families everywhere, the parents of the soccer players had become satellites orbiting their children. Most of them had intentionally kept their families small, so they could devote more time and resources to each child. Almost all of the boys were enrolled in after-school programs for extra help in English or math or science. All of the boys studied at least four languages, reflecting the region's geopolitics: Thai, Burmese, Mandarin, and English.* Under the yoke of demanding parents they were forced to study hard, but they were also rewarded with luxuries many boys in Thailand couldn't afford: Nike soccer cleats, video games, satellite TV, and proper cycling equipment, which for some included clip-in cycling shoes.

---

* Most of the boys also spoke the local Thai dialect, which is as distinct from the Thai spoken in Bangkok as Portuguese is to Spanish. In fact, partly due to its proximity to Laos and Myanmar, there are seven languages spoken in Chiang Rai province.

As thirteen-year-old Queue spoke to Coach Nok on the phone, the boy informed his coach of the plan to bike to the cave. Queue had missed practice that day because he had stayed up late the night before watching the World Cup; however, because he was on the Facebook group with the other team members he knew where the kids were going. Queue said that Coach Ek had taken the boys to the cave on multiple occasions. They loved it. Queue had been on four of those jaunts to the cave in the past year. Typically, he said, they would spend five or six hours exploring, going as fast as they could before turning back. They'd bring the kind of cheap LED flashlights sold in the markets lining Mae Sai's main road. Those flashlights are big business, since power outages are daily occurrences and many locals husband electricity or lack it entirely.

When the boys got as far as they dared, sometimes they'd scribble their names on the walls. The cave was the domain of bats and obsidian blackness, but they were always armed with flashlights, maybe some snacks, and the reassuring presence of Coach Ek.

For some reason, though, after his first trip to the cave, Queue didn't tell his parents about their adventures. Like the rest of the boys, he didn't feel it was necessary. And apparently neither did Coach Ek. Because not only were parents unaware, Coach Nok insists his assistant never told him about the trips either.

It didn't take long for information from boys like Queue to filter through family members and up to Coach Nok: the boys had ridden their bikes to the cave.

The families knew where it was. Everyone did. There's not a whole lot to do in Mae Sai. There are no movie theaters, there's no mall (though there is a giant Tesco store—the British version of a Walmart Superstore). So folks would sometimes head up the dirt track winding past the Sleeping Princess's soaring limestone skirt toward the cave. Once a month or so, Boy Scouts from around the world would camp in a manicured field a few yards from the trailhead that led up to the cave. On weekends a handful of families from all over Chiang Rai province would picnic in a grassy area shaded by the canopy of towering dipterocarp oak trees. The far-off scent of sandalwood—burned as an offering for the Sleeping Princess—mingled with the smell of the visitors' spicy "krap kao" pork. But the poorer farmers who lived in the flimsy, tin-roofed concrete homes in the villages that line the route to the cave would rarely if ever go much beyond the cave's threshold. Local folklore discouraged such treks—it wasn't worth mingling with the cave's spirits, for whom the community had built and maintained a series of shrines just outside the cave. Anyway, they didn't need to go in. The cave site's ancient oaks and outside trails provided natural air-conditioning, beauty, and, for those who sought it, privacy.

On June 23, though, those picnic areas weren't quiet; instead, there was a jet-engine roar of rain—it was the start of monsoon season. Fat drops pinged off corrugated roofs, creating a deafening world of white noise that forces people outside to shout at each other in order to be heard.

The monsoons are a part of life in Southeast Asia, and everyone who has lived through a wet season there knows what to expect. Monsoon systems build in the equatorial Pacific in late April and May; by June they begin crashing into Southeast Asia regularly. Tropical low-pressure systems can squat over the region for weeks. But the rain, especially from May through July, is not constant. The sun often pokes through, squeezing out sky-spanning rainbows. The rains provide a respite from the sapping heat, but more important, much of the local agriculture depends on the wet season—particularly the rice farmers whose paddies line Route 1 and the fields beyond. In late summer you see hunched figures mechanically dipping and planting the next crop.

However, the 2018 rainfall through mid-June had been far above average. In fact, according to the Thai Meteorological Department, Thailand had been pelted with about three feet of rain in the early part of 2018, more than a foot higher than the average rainfall for that time of year. And just days before the boys went into the cave, a low-pressure system stalled over parts of Vietnam, dumping "heavy to very heavy rain," according to the meteorological records. Those rains failed to trigger the flash floods that are so typical in that part of Thailand—when streets and villages are swallowed by water that in a day or so always seems to just disappear—so nobody paid much attention. The boys could not have known that the heavier-than-usual rain soaking the Sleeping Princess in the previous months had caused the mountain to become more waterlogged than normal for that time of year.

The rains made a cave that was typically accessible to amateur cavers in the dry season completely impenetrable in the wet season. The cave complex is a series of tunnels connected by a main channel in the rough shape of a letter T. By monsoon season, the channel that runs through the center of the cave's main chamber is a swollen river pumping out millions of gallons a day, backing up and flooding the main chamber.

The cave was first officially surveyed in 1986 and

1987 by two Frenchmen. The survey ends shortly after the Underwater City, and is punctuated by question marks denoting places that remained to be explored. In 2013, British caver Vern Unsworth plunged into the cave to improve the survey, unaware of the central role he would play five years later in the drama of the trapped boys. During his 2013 survey, he added some annotations, but not many. One of the tantalizing details that British caving guru Martin Ellis would note about this cave in his 2008 survey of Thailand's longest and deepest caves was the existence of vertical shafts leading hundreds of yards upward from the main cave tunnel straight up to the forest above. And in a sentence that years later might have helped launch one thousand troops on a search deep into the jungle above the cave, his paper says: "Elsewhere in the system several high-level passages remain to be looked at. The existence of narrow avens [vertical shafts] opening into the main passage suggest the necessity of surface prospecting as the depth potential is over 600m. The strong air current indicates that there is a link with the surface."

A link to the surface that no one had been able to find.

The calls kept coming. Parents were quickly comparing notes. None had heard from their boys for many hours. At around 8 P.M. Nok got a call from the park ranger in charge of Tham Luang. Damrong Hangpak-deeneeyom had been the head ranger at the cave for two years. The rangers used to take tourists on quick jaunts into the cave—never more than half a mile or so, about a tenth of its total length.

That evening, as the parents were piecing together what happened to their boys, the head ranger got a call from one of his deputies. The deputy park ranger had found a Honda scooter on the road next to the cave and eleven bicycles stashed nearby. Since locals know not to be in the cave after dark, he figured it must be a group of tourists who had gone into the cave, so he decided to scope it out. The deputy grew more nervous as the sun dipped behind the Sleeping Princess. Maybe the tourists were lost. He went in alone, but quickly noticed the rising water. He rushed out to call his boss.

By 6 P.M., the head ranger had arrived with another deputy. The three went in together, their light shoes squelching into the mud. They were armed with older flashlights that barely lit the hazards.

This was technically "his" cave, but the head ranger had only been on station a couple of years, and lacking proper caving equipment, he was terrified of the glowering cave with its now frothing water. At fifty-one, he was the youngest of the trio, and looked it—with a carefully cultivated swoop of black hair and a plump midsection. They made good time to the third chamber, and squirmed past a nasty chokepoint that led into a 150-yard crawl which opens up into Chamber Four—big enough to fit a semitruck. And there on a sandbank they found what would trigger one of the biggest search-and-rescue operations in history: first, a pair of neatly placed black sneakers, then ten pairs of Adidas-style slides and several mud-caked backpacks. Inside the bags they found shirts from the Prasitsart school and soccer cleats.

From the shirts and the size of the cleats they now understood that the bikes and scooter outside belonged to a group of children. The first ranger was now angry with himself for not having inspected the backpacks next to the bikes outside the cave earlier—had he known they were looking for local children they might have come better equipped. Also inside the backpacks were phones and Moo Pa soccer jerseys. They snapped pictures of the backpacks as proof of their find.

With renewed urgency, the three had made for the big T-junction only a few hundred yards on—locals call it Sam Yek. This T-junction has a central bowl-shaped beachlike area. But now that bowl, with the rough dimensions of a swimming pool, was completely filled.

Water was rushing in on the right from the Monk's Series. They assumed the boys must have gone to the left toward the Underwater City, but the entrance was completely blocked by a churning cauldron of water. That meant the boys must be trapped somewhere beyond that submerged tunnel. The water was only waist high, but moving fast, and since they had only been this far into the cave on a couple of occasions over their entire careers, they agreed they'd gathered enough evidence and made a hasty exit.

Mae Sai is a small place. Head Ranger Damrong knew the head coach, so he called Coach Nok and then alerted the local police and district officials. By the time the rangers got out of the cave, a few of the parents and Coach Nok were already making their way up there anyway. They had half jogged into the deepening slop of mud, and it didn't take long to find the evidence they dreaded: the boys' bikes and soccer gear, parked right in front of the cave. But no boys. Dom's mother saw her son's soccer bag and bike and cried into the night, "My heart is gone!"

"At this point I was stunned and shocked," Coach Nok said. "Because I know that cave, and if the water starts rising the entrance [to the cave] is going to get completely blocked off." Coach Nok and the parents ran into the cave, calling out the children's names. The park rangers, alert to the danger, begged them to go no farther. So they shouted into the cave's entrance:

"Night!"

"Biw!"

"Dom!"

"Titan!"

The only answer came from the cave itself, the echoes bouncing the names back at them. The park ranger and the coach tried to console them, as did volunteers who began arriving from nearby villages—who told them, "Don't worry, no one can get lost in that cave." And it seemed true—no one could remember a single incident in which a local had been trapped there. The rangers added that the only person they had ever known to get lost was a Chinese tourist who had parked his bike outside another cave in the park and left it there for two months. After multiple searches they found no evidence of the man and assumed he had gone into the cave and died in one of its distant corners. Two months later a local Muslim charity called about the man. He had apparently spent weeks meditating at the top of the mountain, but would come down every once in a while to buy provisions at a local 7-Eleven. When his money ran out, he sought free meals at the Muslim charity.

So, explained the rangers reassuringly, even this

"crazy Chinese tourist" hadn't really been lost in the cave. They were sure they'd find the boys soon. The ranger and his little team left the parents and hustled to town to retrieve boots, flashlights, and a water pump from their headquarters. When they returned, more parents had arrived. None had left. Many of the parents would not leave that spot for days, refusing to sleep or eat. The ranger soothed them as they mumbled, "How can I sleep when my son is inside?" They couldn't, so they stood out in the rain at the mouth of the cave.

Coach Nok, despite his fear of the place, went back into the cave alone. The rains had slicked everything, and in his sandals he wiped out, flopping on his back and wrenching his neck. He couldn't move, yelled for help, and had to be carried out of the cave with a neck injury that would plague him for months. It was an ominous start to the search.

Chapter Four
Retreat

It was midnight. Governor Narongsak Osatanakorn reluctantly picked up the phone. He was already jumpy after reportedly getting on the wrong end of the wrath of Thailand's ruling junta. He'd only been on the job for a year, but had ordered investigations into allegations of budgetary irregularities and possible graft in local public works projects. He had apparently alleged that the lion's share of money earmarked for his own province of Chiang Rai was instead going to central government coffers in Bangkok. The bespectacled governor was a stickler for the rules, with multiple degrees in engineering. That kind of crusade against corruption comes with major risks in a country run by a military junta.

In 2014, with the blessing of Thailand's powerful

King, General Prayut Chan-o-cha—then commanding The Royal Thai Army—had toppled a caretaker
government to put an end to political violence that
had paralyzed the country for months. Prayut installed
himself as Prime Minister and established a junta called
the National Council for Peace and Security to run the
nation. It was the country's first coup in more than
eighty years. The next six months brought crackdowns
against dissidents and political opponents. Order, if not
democracy, had been restored.

The fuss Narongsak had made and the embarrassment he reportedly caused the junta had resulted
in an impending transfer to a backwater province—
scheduled for just five days hence—from his cushy job
as governor of Chiang Rai.

Now he bolted from bed. "They're what?"

His secretary explained that twelve boys from the
town of Mae Sai and their soccer coach had disappeared earlier on Saturday into a cave and never come
out. Their parents were freaking out; they were waiting at the mouth of the cave. No one had ventured deep
inside. They were waiting for a proper rescue team.

Narongsak, as he's known, hustled out of bed and
sped the forty-five minutes to the cave. The blacked-
out little hamlets lining Route 1 were a flash of graphite
gray and spectral green. He arrived at the cave around

one in the morning. About fifty people were there, a minicamp that over the coming days would swell into a small city. Narongsak was briefed by the head park ranger, Damrong, and quickly organized a rescue team of twenty-two men—local district officials, police, park rangers, and volunteers. They scrambled in shortly after 1 A.M. and reemerged almost three hours later with grim news: the rescue team had made it to the T-junction, and not only was there no sign of the boys, there was a tremendous amount of water—as powerful as an ocean rip current. And it was rising.

They analyzed the situation: The tunnels beyond the T-junction seemed completely submerged. The terrain that was not submerged was merciless, with not a single patch of flat ground. The darkness was disorienting, there were trip hazards everywhere, and there was zero communication to the outside world. Many of the passages required crawling—so forget about ambulance gurneys or even regular stretchers. Cell phones had no signal, and because rock is conductive, the cave swallowed up the radio waves of their walkie-talkies. It was the worst possible place for a search-and-rescue operation, and Narongsak, ever careful, wanted to avoid a situation in which rescuers would need to be rescued.

So Narongsak ordered everyone to pull back. They

needed to reassess, and he needed someone who actually knew the cave.

Early the next morning he convened another meeting. At 9 A.M. on Sunday, June 24, rescuers—including a sixty-three-year-old foreigner—started trickling into the cave site.

No one knew it yet, but his presence was the first bit of luck the rescue would encounter. The foreigner was a sinewy Brit named Vernon Unsworth, known to everyone as Vern. Vern, a ranger told Narongsak, was the crazy Brit who went into the cave all the time.

Vern had fallen in love with caving at sixteen, joining the Red Rose Pothole Club near his native Lancaster. He just happened to be born on the doorstep of the United Kingdom's caving country. Lancaster is at the stem of what on a map looks like a giant cloverleaf of three national parks, featuring some of the country's longest and deepest caves. On weekends young Vern would explore the fifty-three-mile-long Three Counties cave system nearby. Still in his teens, he'd climbed the Matterhorn and Mont Blanc—in the same week. He took a decade-long hiatus to race motorcycles semiprofessionally, then returned to his first love, caving.

Divorced after two decades of marriage, he had recently fallen in love with a local woman who lived in Mae Sai. Her name was Tik. His Thai was rudimentary, so he spoke to her in the singsong-accented English that betrays his roots in northern England. From the window of their little bungalow, he could see the entirety of the Sleeping Princess—Doi Nang Non mountain.

Vern loves Tik, but as he tells it, the object of his deepest desire was that Princess—more precisely, her guts. He'd been exploring the Tham Luang cave for a few years now. In 2013 he and his caving buddy, sixty-five-year-old Rob Harper, started redrawing the original 1980s survey map. That map had charted the cave as being about four miles long. In Tham Luang, Vern and Harper had taken a right at the T-junction and pushed forward up through the Monk's Series, the tunnel and its cramped offshoots that point due north. The shafts were so tight and the belly crawls so gnarly that Vern's back looked like it had been shredded by a feral cat. He took a selfie of his injuries, which he likes to show friends—or reporters—when they ask about caving. But that expedition added over half a mile to the cave's overall documented length and Vern decided to explore it again the next year.

The locals, he says, "thought I was crazy. For the last four or five years, they called me Crazy Caveman. They couldn't understand why I spent so much time in caves. Caves to them are supposed to have a spiritual side. They could not understand why I spent so much time there. What's the point?"

If you had asked that question to Sir Edmund Hillary and Tenzing Norgay about Mount Everest, or to explorers like George Mallory and Robert Falcon Scott—who died pursuing similar obsessions—they might have answered, "It was there." But "there" also encompasses the uniquely human impulse to be first—to explore the blank parts of the map. And cavers submit that the greatest trove of unexplored places on earth is the treasure chest of tunnels and passageways that honeycomb our hills. Plus, unlike seventy-thousand-dollar-per-person Everest expeditions, it's a stunningly egalitarian activity—all you have to do is get there.

For as long as humans have been around, caves have served as places to live or in which to store what was most valuable or most dangerous. Neanderthals crammed their clans into caves also inhabited by various animals. Cro-Magnons famously painted on cave walls. Today deep caves, sometimes man-made, serve as cold storage for the planet's seeds (the Svalbard Global Seed Vault) and as the future depository of America's nuclear waste (Yucca Mountain).

But going into caves for recreation, rather than for shelter or storage, began in the late nineteenth century.

Caving was pioneered in the late 1880s by Édouard-Alfred Martel, who started poking around deep holes in the ground in France. In 1895 he successfully descended into the nearly vertical 350-foot main shaft at Gaping Gill in Yorkshire, England. The quest to go deeper led him to develop his own techniques and the rudiments of modern climbing equipment, including thinner climbing ropes and metallic ladders. A group of Frenchmen, including a metal machinist named Fernand Petzl—with apparently little else to do during World War II—descended into the Dent de Crolles cave system near Grenoble, France. It was during that record-setting expedition—they traveled over two thousand feet into the earth—that primitive prototypes of the modern nylon climbing rope, rope ladders, and rope ascenders used in lieu of ladders were developed. Petzl continued to tinker with climbing safety gear—ultimately founding the eponymous climbing manufacturer Petzl. The descendants of those prototypes are still used by climbers and cavers today.

In the British Isles, recreational caving started during the latter years of the nineteenth century, but until the period between the World Wars it remained largely the preserve of a very small group of adventurers and scientists interested in the unique geology and biology found deep inside the earth. During the 1920s and '30s interest in caving grew, leading to the formation of the first caving clubs, initially in the Yorkshire Dales near Lancaster and in Somerset.

England is notoriously wet, and the north is typically cold *and* wet. By the mid-1930s in England, cavers who kept running into water devised ways of crossing it. In 1936 Jack Sheppard made cave-diving history by crossing a short little "sump"—an important cave-diving term for a submerged section of a cave between two dry passages—in a cave in southern England using a suit filled with air from a modified bicycle pump.

The growth of the sport's popularity worldwide correlated with a commensurate rise in disasters, many of them involving water. In 1993, south of St. Louis, a group from the St. Joseph's Home for Boys set out to Cliff Cave County Park. When heavy rains drenched the area, most of the boys and counselors left the cave, but five boys and two adults stayed to explore further. Rainwater gathered in streamlets, which poured into the cave's sinkholes. Six members of the group drowned, four of them children. One of the boys, thirteen-year-old Gary Mahr, managed to survive by clinging to a rock shelf above the water for eighteen hours.

A similar incident occurred in England in 1967, when a group of some of England's hotshot cavers, led by a swashbuckling twenty-six-year-old named David Adamson, became trapped as waters flooded the Mossdale Caverns in Yorkshire. It was already soggy before they went in. The forecast had called for thunderstorms. Anticipating the flooding everyone suspected possible, and defying the group leader's exhortations to stay put, four of the original group of ten fled the cave, including Adamson's fiancée. Those four survived, but a massive effort failed to rescue the six who stayed. All six young men died in what was then the world's worst caving accident.

Mossdale Caverns happens to be just a few miles from Lancaster, where the twelve-year-old Vernon Unsworth was on the cusp of a lifetime obsession with climbing and caving. Of course he knew about the failed rescue effort, the whole world did. Hundreds of locals, some digging with bare hands, had built a makeshift dam to divert the stream that stuffed the cave's belly full of water. Retching from exhaustion and choked by churned-up foam, the rescuers discovered five of the six bodies jammed up into small passages. It would take two more days to find the sixth member of the doomed expedition, half-buried in mud. The search, the recovery of the bodies, the funerals, and the finger-pointing made headlines for days.

"They were playing Russian roulette, really," Vern said, fifty-one years after the tragedy. "We all thought about it. But you just have to get on with it, don't you?" Mossdale was hardly a glamorous cave; it was basically a miles-long mole burrow, with ominously named sections like Rough Chamber, offering none of the geological delights of caves like Tham Luang. Mossdale's distinct appeal was its length and the possibilities it offered for a connection with another long cave system. Vern and others would explore it countless times over the years, cautionary tales be damned.

"It is the excitement," continues Vern. "Finding passages, finding new areas of the cave where no one has ever been before. Those footprints you make? Those are yours, you're the first. It's a bit like Neil Armstrong on the moon. You are the first person in that part of the cave. No one has ever seen it before."

There's a unique kind of satisfaction in finding new passages or connecting existing cave systems—something primal that certainly did not exist in Vern's day job as a mortgage broker and financial adviser.

Tham Luang was Vern's hobby, and he freely admits that it was quickly turning into his obsession. It's a caver's dream. Located in the Sleeping Princess's head—where her earring might be—it's easily accessible for him off a decently maintained dirt road outside Mae Sai, which he calls home for part of the year. Its internal curtains and kissing pairs of stalactites and stalagmites are protected by a dedicated group of park rangers. Beyond that entrance it is rarely traveled, and its deeper recesses harbor endless unexplored mysteries. There's another convenience for Vern; it's only a few miles from Tik's house. In March 2014, Vern and Harper were back, this time turning left at the T-junction. They reached the end of the cave, past the shrivelingly cold Goolie Cooler to the perennially wet Voute Basse, and hit a dead end.

Caves are cryptic places, where rock is folded all over itself, leaving an often hidden geological wrinkle of a passageway. And every cave has a tell. Like humans and fire, caves need to breathe. A whisper of wind means there's an opening somewhere. The bigger the draft, the bigger the possible opening beyond. And Vern felt, and heard, something. For sixteen hours during the expedition he'd schlepped a five-pound hammer and a chisel in his pack. He felt his way toward the source of the draft, a small "window" beside the water-clogged passage of Voute Basse, and could see a large cavern beyond. He started to hammer away, enlarging the window enough to squeeze through and reach uncharted territory. If you're a caver, this is winning the lottery. It was terra incognita. Cavers have a term for this: "scooping booty," as in grabbing treasure. To them uncharted territory was worth more than gold.

He'd found what cavers call the master cave—the source of water in a particular cave system. In a single expedition the two men had turned Tham Luang from the sixth-longest cave in Thailand to the fourth-longest. It was now over six miles long. And they kept at it, going back in 2015 and 2016 to map new uncharted extensions to this mysterious cave.

Early on that rainy morning of June 24, 2018, Vern committed himself to the growing rescue effort; he called a local caving buddy named Lak, who was a volunteer park ranger. When Lak arrived, they went in.

How could anyone get lost in Tham Luang? Vernon thought to himself. It's a straight shot, and any reasonably fit person could handle the terrain—certainly a youth soccer team could. The only thing you need to remember is whether you took a right or a left at the T-Junction. Like most, he figured he'd fish them out and be done with it. He and Lak cruised through the main chamber on that Sunday morning and scuttled through the pinches leading to Chamber Two. Just past

Chamber Three, Vern noticed water piling up in the gutterlike sides of the tunnel, but he was on autopilot until they reached the T-junction.

It stopped him in his tracks. The bowl that he'd seen so many times was now overflowing. He'd been told there was water, but didn't expect this much. They reckoned that the boys had gone to the left. To the right was the Monk's Series, that uphill wash, and the boys would have been flushed out. So they turned left and began to shout, yelling the boys' names. By then, though, the water had fully flooded the passage ahead, blocking their voices.

Neither man had been in the cave during the wet season, and both found this watery world almost un-recognizable. Vern thought the flooded T-junction was diveable at the time, but no one in that search party had either the know-how or the equipment to do it. They turned back—soaked with sweat and cave water and battered with a layer of the reddish mud. It would be the first time the cave had defeated Vern and the first of a ceaseless series of retreats in the coming days. Still, Vern trudged out of the cave with an idea. If they could block the water flowing down from the Monk's Series, they could arrest the rising flood. After a meeting with Governor Narongsak, they decided to return—this time with a small contingent of local soldiers who had arrived and a team of technical rope rescuers and divers from the nearby city of Chiang Mai.* The leader of that team, Noppadon Uppakham, called Taw, is an experienced caver who works for a rock-climbing outfit run by his brother-in-law, an American living in Chiang Mai named Josh Morris.

Vern went in again, this time with the climbing team dragging in a pump and digging tools. They drew markings on the wall, to more easily quantify the rising water level. It was easy going until that tight space before Chamber Four. To traverse tight passages like that you have a couple of choices: scooting on your butt with your hands behind you (for short distances), which burns your thighs—crab-walking—or crawling on all fours. Either way, after hauling in the gear they were all out of breath by the time they reached the T-junction. Vern informed the team that there was little chance the boys had headed to the right, uphill toward the pulsing waters coming from the Monk's Series. The team's plan was to simultaneously dam the water flowing down with sandbags and pump enough

---

* Not to be confused with Chiang Rai. The confusion is so widespread that a leading American broadcaster mistakenly sent its lead engineer to the wrong city. Nobody could figure out why they couldn't find him at the airport. Eventually he took a four-hour van ride to the right city.

water out to allow the flooded section to be forded. The sandbags didn't need to arrest the flow, just slow it enough to allow the pump to work. The soldiers immediately got out their shovels and started filling sandbags. As soon as they placed one in the water, the current zipped it downstream. Abandoning the sandbags, they would have to rely on the pump alone. They cranked it to life and sent one of Taw's divers to inspect the section on the left side toward Pattaya Beach and the Underwater City.

Even though Vern knew there was a passage to the left, they could not see it. The diver came back after a few minutes, saying he had found only rock walls and mud. Vern wondered if the mud in the water had clogged the entrance to the passage.

The pump ran on a diesel engine. And while it seemed to siphon off a decent amount of water, it was also coughing out a cloud of fumes and carbon monoxide. The men carved little seats into the bank and waited. Minutes later the futility and the fumes drove them all out again.

Like a failed siege of a medieval city, wave after wave of rescuers threw themselves against the cave, its water, and its walls. Increasingly bigger military units were called up for the assault.

Later that evening a contingent of the Thirty-seventh Military District Unit based in nearby Chiang Rai was dispatched to the cave, under the command of Colonel Singhanat Losuya—who would play a key role later on in this saga. But that night, his chief concern was the parents. They were gathered at the mouth of the cave, weeping. As his team went in, they kept asking, "Colonel, can you find our boys? Where are they?"

In the hierarchical culture of Thailand, rank and seniority are strictly observed, and the parents kept a respectful distance from the colonel. As the father of two boys nearly the same age as the boys in the cave, the colonel was moved, but he had no answers for them. His team was threading its way in, and at that point there was no way to communicate with them. Still, the arrival of the military frightened the parents. In the ranger hut where they fled from the rains they asked why they needed the military, was it that bad inside? Gently, the rangers told them the water was certainly something they could not contend with given the limitations of their training, and assured them the soldiers of the 37th would find their boys.

The colonel's first team in was comprised of nine soldiers and one officer—Captain Padcharapon Sukpang. This officer also had two district officials and a park ranger with him. The captain, a career soldier of over twenty years, kept a diary and faithfully logged his missions. He said his little group had no trouble making it to Chamber Three. It was there that they began to notice the water rising everywhere.

They forged onward. The captain is in his fifties, but remains an athlete. He has a high-and-tight haircut and wears an even tighter uniform, even when the heat makes it hard to breathe in Mae Sai. But he didn't mind getting that uniform dirty when it mattered. Which is good, because by the time they exited Chamber Three and began the 150-yard crawl onward, he was lathered in mud.

The water was now seeping from the ceiling, as if the cave had broken out in a massive sweat. It was pooling on the floors, which dip lower on the sides, like a natural drainage ditch. When the group was a few yards away from the T-junction, they began smelling the fumes from Vern and the climbing team's aborted pumping operation. The captain noticed a sandbag near the T-junction's bowl and thought that using it was like trying to use a pebble to dam a river. One of his men waded out into the water—now about five feet deep. They also didn't have the proper equipment so they turned back (a trend, at that point in the search).

But before heading out they figured they could at least measure the speed of the water. They filled a water bottle a third of the way, capped it, and chucked it into the stream. They calculated it sped south toward Pattaya Beach at about twenty miles an hour, a startling speed given that they were inside a cave. And as they turned around to leave the captain spotted something. Names on the muddy cave wall. He showed me pictures of them. Some appeared to have been drawn with a finger in the mud-encrusted walls. Others seemed to have been carved with a tool or stick. He believed they matched the names of the missing boys—one of the names still legible was Pong. There was also an arrow, he said, pointing south toward Pattaya Beach.* The captain rushed out of the cave to inform his commander. It played big in the local papers the next morning, when a trickle of reporters and local TV outlets started to snoop around, wondering where the boys were.

There were now a hundred people populating the little campground outside the mouth of the cave—many of them wearing thin, brightly colored plastic ponchos against the rain. The picnic area's grass started giving way to mud under the ceaseless rain and the countless

---

* He did not take a picture of the arrow, and in retrospect it makes little sense that the boys would have drawn it as a distress signal at the T-junction: on the way in they would not have been under duress, and had they made it there on the return trip they could have just walked out of the cave.

footsteps of nervous parents, volunteers, and newly arriving troops. Noting the repeated failures of the rescuers, the parents started making offerings to the cave spirits—incense and fruit juices. The spirit of the Sleeping Princess had been dormant for so long that no one could understand if or why the spirit had become angry. Finally the parents succumbed to fatigue and the entreaties of the park rangers. They collapsed onto thin mats on the floor of one of the ranger huts and tried to sleep as the mosquitos feasted on them.

As Vern left the cave complex that night of June 24, he had two distinct realizations. The first was that his beloved cave had turned on him. It had become a monster that repulsed even its most faithful benefactor. The second was equally troubling: *It could have happened to me.* The Saturday the boys disappeared he'd extended his visa in order to explore the cave. He had planned to go in on Sunday, June 24—which he did, but as a rescuer rather than as a recreational caver. And because of the way the cave is configured, if one is heading for its far southern reaches there's no way to hear or see the sickening surge of water from the Monk's Series down to the T-junction. You wouldn't know it until it cut you off on your way back. That's what happened to the boys. And even as a seasoned caver, it could have happened to him.

"The timing," he said, "was just incredibly unlucky. It could have happened to me," he repeated. But it didn't. He was out. They were in. And clearly they needed help. At a meeting that night Narongsak decided to call in the heavy guns.

Chapter Five
"You have one last chance, or the boys will die"

It was becoming a trend: high-ranking officials roused late at night. It was Asahna Bucha weekend in Thailand, a national holiday commemorating the founding of Buddhism. Observed on the first full moon in the eighth lunar month, it commemorates a sermon the Buddha gave twenty-five hundred years ago. It's a long weekend, and that Sunday night, June 24, Rear Admiral Apakorn Yuukongkaew was settling down to sleep at his base near (the real) Pattaya Beach when he got a call. His Thai Navy SEALs were needed.

"Send our team to help the boys, tonight," the commander of his battle squadron, a full admiral, ordered.

Apakorn, a Thai Navy SEAL vet with a tight crew cut and a hangdog countenance, had heard about the boys but hadn't paid much attention to the story. He was the kind of leader who spoke to his men rather than shouted at them, whose pre-mission briefings were matter-of-fact, devoid of the rousing rhetoric of a Patton or a MacArthur. The men respected him because he was their commander, their senior, and because he respected them. He had the self-assurance to delegate to his junior commanders, allowing them to make tactical decisions in the field.

By now it had been more than twenty-four hours since the search began. The news had started to make waves in the Thai press, and over the next twenty-four hours would be beamed across the world via a short bulletin on CNN. The Associated Press had started covering the story, sending out a few slim dispatches, as journalists started to book local hotel rooms. Satellite trucks were now parked alongside military pickups. Canopy tents to house the reporters went up and folding tables were placed in the mud. Government officials, including Governor Narongsak, were telling reporters that they believed the boys were alive. But it was just a guess, based only on hope, not on a single fact. Pictures published in the *Bangkok Post* and *The Nation* (Thailand) showed teams of uniformed police and volunteers in bright orange reflective vests lining the entrance to the cave, which already looked like an elaborate movie set. Panel lights lit up the cave walls as if for a massive museum restoration. One photograph showed Governor Narongsak, looking out of place in the land of rock and mud, hands in his pockets, staring down. He wore suit slacks, a crisp white shirt, black lace-up oxfords, and the unmistakable look of defeat.

It was Narongsak's decision to call in reinforcements. By kicking it up the chain he elevated the matter to a national level. Now the head of the navy was involved, and by the time the sun warmed the Sleeping Princess's forests the next day members of the nation's cabinet would be alerted. Narongsak was nominally the incident commander, but he was no longer in charge. Nor, for that matter, would Apakorn be in charge of the military effort. The commander of the Thai Third Army would essentially be in command, and soon enough even he would answer to higher powers.

Nevertheless, the distinct sense of urgency presented an opportunity for the Thai Navy SEALs to dip in and snatch some glory. Founded in 1953 (reportedly with help from the CIA) the Thai Navy SEALs maintain close ties to and even train with the American Navy SEALs. Like their American counterparts, they are their country's elite fighting force. Each Thai SEAL undergoes a six-month training course that includes the Thai version of "hell week," in which SEALs are subjected to torturous group challenges in order to weed out all but the toughest and most dedicated candidates. And given that Thailand's two-thousand-mile coast-line wraps around a big chunk of Southeast Asia, they are frequently deployed. With their work in counterinsurgency, antipiracy, and high-profile rescues, Apakorn thought it was "not a difficult mission." After all, they were the kind of badasses who lived hard, smoked hard, and won trophy wives. But the orders came from on high, and Apakorn didn't question them.

By eleven thirty Sunday night, a spearhead of Thai SEALs began loading the unit's Embraer jet. On the tarmac around the muscled men were dozens of hard-plastic Pelican cases and dive tanks. At 4 A.M. on Monday morning, June 25, a Thai SEAL detachment became the first wave of about twenty Thai Special Forces troops on the scene. After initially scoping out the cave, the team went in. The mission was nearly suicidal, but they managed to dive across the T-Junction and make it nearly to Pattaya Beach—farther in than any of the rescue teams thus far.

Cave divers use a guideline to direct them. The line, typically a rope not much wider than a laptop cable, is strung along the route in a cave. It tells you where you've been and where you're going. The first diver to an area spools out the line as he goes, weighing it down with small sandbags or tying it off and generally keeping right down the center of the passage. But that first contingent simply wrapped ropes around their waists and dove in. In a cave spiked with glass-sharp stalactites, a rope like that could easily snag or be severed. And with zero visibility, it's an expedient way to get killed. But the initial troops hadn't been trained in cave rescues or even cave diving. They had no cave-diving kit, which includes multiple air tanks (side-mounted tanks are used to slip through cave choke points without having to remove a tank from one's back), proper guidelines, and digital compasses. They had no maps or surveys. What they possessed aplenty, though, was courage.

According to the accounts of the Thai military, once the SEALs crawled up on Pattaya Beach and caught their breaths, they noticed footprints. Their commander says they saw a handprint on the wall. *It must have been the boys*, they thought. They had also apparently found a length of green rope, the same rope Coach Ek had used. By this time, the commandos had consumed most of the air in their tanks. Still they forged ahead. The Thai SEALs couldn't have known it, but they had just passed the spot where the boys and the coach had spent the previous night.*

---

* Some of the foreign teams that would later arrive at the search-and-rescue dispute that they ever made it that far.

Scuba diving in the open ocean is relatively straight-forward—if things go wrong, your escape hatch is generally straight up and as wide as the horizon. In contrast, cave diving can be deadly because your escape hatch might be miles behind you. Straight up—or, for that matter, left, right, down, or sideways—is rock. Cave diving was pioneered by the British ducking into those goolie-shriveling sumps in northern England. Only a tiny fraction of people choose a hobby that can involve such inherent discomfort and risk, yet over the past ninety years more than 130 people from Britain alone have died cave diving. That's a big number for a small group. The following is a grim sampling of the ways they died: crushed by falling rock, hypothermia, drowning, natural causes (e.g., heart attack), falls, asphyxiation, CO_2 poisoning. Many of the casualties' bodies weren't recovered until years later.

Because of this heightened danger, there are five basic rules of cave diving:

1. Never dive beyond your certification level or your technical capacity.

2. The rule of thirds—never use more than a third of your breathing gas on the way in—you will

need a third to get out and the last third as a
reserve.

3. Maintain a physical guideline to the entrance of
a cave.

4. Never dive below the depth of your breathing
mixture—a cocktail of gases that help prevent
divers from getting "the bends" or decompres-
sion sickness, a condition that occurs when divers
ascend from a depth and gases build up in their
tissues. It can be extremely painful and potentially lethal.

5. Carry at least three lights per person.

Out of necessity and innocent ignorance of these
rules, the SEALs violated all but rule number four,
which also happens to be the only one that doesn't
apply in Tham Luang; they didn't have to worry about
the gas mixture because this cave had no deep dives
and therefore no bubble-causing ascents.

Once they crossed the dry section of Pattaya Beach,
the Thai SEALs were now staring at the cave's longest
submerged passage: over three hundred yards of snags,
rocks, and a tar-black river of water. Not only did they
not have the right equipment, they didn't have a single

map or survey—possession of which should perhaps be an unofficial sixth rule of thumb. Out of gas and pretty much everything else, they too turned back.

Amid the mud-spattered crowd watching the Thai SEALs stumble out of the cave that rainy Monday evening were Vern Unsworth and Ruengrit Changkwanyuen, a compact forty-two-year-old with a hint of a goatee. Ruengrit is a regional manager for General Motors, based in Bangkok. But he is also one of the most experienced cave divers in a country where caving is mostly the domain of tourists. Like millions of Thais, he'd been glued to reports from Tham Luang. Unlike the millions watching on TV earlier that Monday morning, he wasn't so much proud of that first contingent of Thai Special Forces that exited the cave as he was terrified for them. Watching from his home in Bangkok, he needed only one look at their equipment—the single tanks, the wet suits designed for combat in tropical water—to know they were in mortal danger. Within an hour he called his boss to tell him he was off to Mae Sai to help with the rescue—he reckoned he'd be gone only three days.

As it turned out, Ruengrit possessed two skills that would prove invaluable. One derived from the other: he'd spent many years in Michigan, where he picked up fluent English and where he also learned to dive.

That Monday evening, when he spoke to those Thai SEALs coming out he was humbled by their exploits and aghast at the risks they'd taken.

"They were clearly willing to die to find these boys," he recalled. "No matter what, they needed to find the boys."

That night—Monday, June 25—he told the Thai SEAL shift commander he could help. By six o'clock the next morning, he'd hauled up all of his personal equipment and given the Thai SEAL dive teams a crash course on equipment and technique. He was impressed by how quickly they learned. The commander's eyes bulged once he saw Ruengrit's gear—his side-mounted tanks, his helmet studded with lights, the rope he used—and immediately put in a call to the Thai SEAL base in Pattaya for side-mounted tanks, more regulators, caving helmets, and harnesses.

By late morning on Tuesday, June 26, the Thai SEALs whom Ruengrit had briefed had hardly become cave divers, but at least they were no longer a suicide squad. He led what was now about the ninth search mission as they smoothly traveled all the way to the T-junction, where they found foaming, Guinness-colored water. It had been raining at the rate of a couple of inches an hour, off and on, for more than two days. By now, even-muddier water had begun surging up from the southern end of the cave—the direction of Pattaya Beach—and was slamming against the clearer water coming down from the Monk's Series. It looked like a miniature version of the waters off Cape Horn, where the Atlantic and the Pacific crash together. Even with Ruengrit's skill and gear, they couldn't pierce it.

Instead they started taking measurements of the water level and tried to string out a guideline. But the froth was rising by the minute. Within three hours they'd been forced to back up three hundred yards. The current now pulsing toward the entrance of the cave was unstoppable. The only option was retreat.

With the SEALs' diving operation failing to make headway, attention turned to the idea of pumping the water out—it mattered little where the water was channeled to, as long as it was out of the cave and downhill. Governor Narongsak had already initialized plans to begin pumping water out of the cave—early on he had declared water the enemy—but total victory wasn't necessary. All that was needed was a stalemate that could stabilize water levels and reduce the current that the divers had to fight against. With that in mind, on Tuesday a bright-orange industrial submersible pump had been delivered, along with miles of thigh-thick hose to siphon the water down the hill from the burgeoning rescue camp. Maddeningly, it arrived missing crucial parts and would sit there unused for another twenty-four hours. There were some other, smaller pumps going, but they did not seem to make a dent. With the near-constant rain, oil-slick mud, broken pumps, and impenetrable wall of water, things were looking grim that Tuesday afternoon.

Through his caving circles, Vern knew about a pair of crack British cave divers, Rick Stanton and John Vollanthen, who had led multiple cave-diving rescues over the past decade—something no one presently in Thailand had ever done. Over the previous twenty-four hours, Vern had watched the water conquer more and more of the cave, and he began quietly agitating for an international crew of scuba divers with enough skill to survive this cave. He had secretly called his caving buddy Rob Harper to put Stanton and Vollanthen on standby.

By the late afternoon of Tuesday, June 26, Vern felt the time for quiet agitation was over. He buttonholed Governor Narongsak in the park ranger station, which had turned into the search's headquarters. He told him the only way to find the boys, dead or alive, was to send in divers—the best divers.

"You have one last chance, or the boys will die," the excitable Brit told the governor. "You have to call in these divers, now." But Vern says Narongsak seemed to ignore him, and walked off to a meeting.

Vern knew that he had offended Governor Narongsak. He also knew that he'd lost face himself when he lectured the senior politician with multiple advanced degrees.

As it turned out, Stanton and Vollanthen already knew all about the ongoing rescue before they got the call from Harper.

Just weeks earlier, Stanton had met a woman named Amp. The thirty-seven-year-old had been a nurse at an assisted living facility in Chiang Mai, Thailand, which caters to foreigners—the care there is said to be excellent, and far less expensive compared to similar properties in England. The elderly couple Amp had worked with happened to be the parents of one of Stanton's old caving buddies. In late May of 2018, Amp visited England, staying with the family in the south of England. She told Stanton's friend she was interested in kayaking, so he asked Stanton—as avid a kayaker as he was a caver—to scope something out for them. Amp is petite, with a high forehead and bubbly energy. Stanton, fifty-seven, is not a particularly effusive person, but he describes being "quite taken" with her—which for him is akin to gushing. Amp worked in Chiang

Mai, but in an uncanny coincidence happens to live in Chiang Rai, Governor Narongsak's provincial seat and less than an hour from the Tham Luang Cave.

Amp happened to be visiting Stanton in late June when the boys went missing, and early on June 24 she excitedly brought him the news bulletins being cranked out by her hometown news sites about a soccer team trapped in a cave. Stanton's first call was to his partner Vollanthen. Stanton had limited his participation in big cave-diving expeditions in recent years as the muling of heavy gear and the logistics had become tiresome. He now preferred open-water kayaking. So he was ambivalent when he got Vollanthen on the line.

"John," Stanton asked, "is this something we are going to get involved with or shall we sit and wait?" Vollanthen, nearly a decade younger than Stanton and a father, voted for immediate action. In the end, the decision was moot, because the action soon came to them.

The clock was ticking, not only for the boys inside the cave but also for the British divers, who by now had been on standby for twenty-four hours. It was vacation season in England, and Stanton and Vollanthen were about to go off the grid. Rob Harper, Vern's caving mate and his main liaison to the rescue divers, was also about to head out for a holiday with his wife. On Tuesday night, with the rain pummeling the little park ranger station, Vern sat in on a meeting with Interior Minister Anupong Paochinda and Minister of Tourism and Sports Weerasak Kowsurat. It was around 9 P.M. in Thailand, 3 P.M. in London.

According to Vern, he slid a note with Stanton and Vollanthen's names and contact information across the chipped folding tables. The tourism minister opened it and asked, according to Vern, "What do you want me to do?"

Vern, whose voice tends to rise and crack when he gets excited, blurted out, "What do you *do*? Call them now. Here's my phone. Call Rob."

He had already opened the WhatsApp app to Rob Harper's contact when he handed the phone over to the minister. Either the minister or Vern must have pressed the video call icon instead of the phone icon. The tension built as the phone rang and rang. Finally Rob Harper groggily answered—in his pajamas, lying in bed. The minister was taken aback. Wasn't it the middle of the day in London? Why is this man, upon whom the boys' fate might rest, in pajamas in the middle of the day? Harper explained that he is a veterinary surgeon who often works nights and sleeps in the day. Now he was fully awake.

Within three hours he, Stanton, and Vollanthen would be on a flight to Bangkok.

In a matter of hours, the boys had gone from a side-bar news item to a national obsession. The boys. The cave. The rescue. The pumps. The divers. The Thai SEALs. The banner headline in the *Bangkok Post* read HUNT GOES ON FOR KIDS IN CAVE. Narongsak, just two days ago facing the rapid demise of his political career, was now holding multiple daily press conferences that Thai TV ran live and uninterrupted.

By Wednesday, June 27, there were about one thousand troops and rescuers at the cave. No expense was spared. Equipment started pouring in. It looked like an outdoor concert venue. Bundles of electrical wires, hoses, and pipes snaked into the cave—the mouth of which now glowed with stadium lighting. The cave complex's parking lot had turned into an auxiliary camp that looked like Woodstock, with a parade of mud-washed people passing full kitchens whipping up hundreds of meals a day, a medic's tent, lights, and barricades. The Wi-Fi routers and extra cell phone tower provided some of the fastest and most reliable connectivity in Thailand. (Bathrooms, however, would remain scarce throughout the entirety of the search-and-rescue mission.)

Rescuers were sleeping in their cars. Soldiers flopped down at the cave's sandy entrance, sleeping in their fatigues, so exhausted that neither the lights nor the racket made by the next shift disturbed them. The Thai Navy SEALs' Embraer jets started to shuttle between their base outside Pattaya and Chiang Rai's airport, carrying gleaming new air tanks and hundreds of Pelican cases cradling high-tech gear, including snaking videoscopes to peer into inaccessible creases of the cave, sonar, and radios. All of it was stacked neatly onto trucks for the forty-five-minute run to the cave.

That Monday, June 25, an Israeli living in Bangkok had received one of the thousands of calls made to anybody in Thailand with the skill or connections to help the boys. Asaf Zmirly had been marketing Israeli high tech in Southeast Asia for the past eight years, specializing in military-grade emergency equipment. He had a load of contacts, one of them a contractor to the Thai Navy. He'd tried to sell them devices that would detect survivors trapped in rubble—a system that saved lives in the Mexico City earthquake of late 2017. The navy contact told him they needed durable communication devices that would work in a cave and could get banged around a bit: Did Zmirly have anything like that? He did: one of his clients was an Israeli startup called Maxtech Networks; Maxtech had developed a system of handheld radios featuring a "daisy chain" of wireless technology that—much like the Internet—employs a sophisticated algorithm to route data and voice communication via the best source available. It's basically a robust, mobile Wi-Fi system.

Zmirly contacted the company to ask if they could sell and ship the product that very day. When company CEO Uzi Hanuni learned where it was going, he offered the system free of charge, loaded a technician with seventeen of the devices, and packed him off from Tel Aviv's airport to Bangkok. The technician arrived on Tuesday, June 26. But Zmirly and the tech encountered the same obstacle that had ground the mission to a stop: the water.

The system worked well in the dry sections of the cave, but it didn't work underwater, so Zmirly and the tech set off to the open-air bazaar that spills out from the Thailand border into Myanmar. They needed a long cable to connect units on either side of the sump at Chamber Two. Their first stop was at an internet and cable TV provider, where they grabbed the longest data cable they could find, a fifty-yard-long coaxial cable. At another store they found a soldering device, and at a third store the necessary waterproofing. During their three-hour shopping expedition, the Israeli pair never had to open their wallets: the Thai proprietors noticed the dried mud caked on their clothes, realized they were part of the search mission, and refused all payment. As they waited for the waterproofing in that third store, the foreigners were even brought savory rice bowls for lunch. "It's impossible to overstate how generous everyone was. Every single person we met in Mae Sai was desperate to help in any way."

They hustled back to camp and began tinkering with the soldering device and the cables. Finally they had something they hoped might work. Maxtech's lead engineer in Israel warned them that the system wasn't designed to accommodate the kit they had MacGyvered, and predicted that it would fail. "Well, we dove into the second sump to the third chamber, turned on the devices, waited, and you know what, it worked! On the very first try." The system became the main link from Chamber Three to the outside for days to come.

The search-and-rescue mission now boasted some of the world's best cave divers, hundreds of crack troops, world-class technology, and a bottomless supply chain. At their meeting that night, everyone agreed with Narongsak that water was the enemy. What they seemed to lack, according to military officials and foreign experts on the scene, was a coherent strategy. Arguably the meddling of cabinet ministers, top-ranking officers, and even members of the royal court—each with their own preferred mode of action—bogged down the process, creating a hydra-headed chain of command whose orders were nearly impossible to divine, much less follow.

And in the corner of the camp, downhill from the headquarters where the generals and politicians sat, just off from the overflowing row of toilets, were the parents. Like everyone, they initially believed the boys would be found quickly and brought out to them—maybe in a day or perhaps two. There was now no proof their children were even alive. So again they beseeched powers a few rungs up from the ministers: the spirits of the cave. This time they presented them with a lavish meal set on a picnic blanket: fresh coconuts (with a plastic straw for the deity's convenience), mangoes, papayas, lychees, melons, rice cakes, sodas, slushies (also with a straw), and several wreaths of orange marigold. They knelt there in the rain, on the edge of the picnic blanket, praying to the Sleeping Princess, the one who had killed herself and her own unborn child, to spare theirs.

Coach Ek had promised the boys the waters would recede, but they hadn't. In fact it had been the opposite.

On their first night in the cave they'd made camp at Pattaya Beach; the water had woken them up and forced them to retreat farther away from the cave entrance. The second night, the water had pushed them back again. And finally they found ground high enough above the canal below to keep them dry—it would later be called Chamber Nine. They could not have known it, but this was likely the sole place in the cave along the multi-mile route past the T-junction that would stay dry throughout the monsoon season.

If you could peer through the ropy canopy of bamboo and towering oak, through the ocher-stained sediment and the quarter mile of limestone that served as their roof and possible tomb, you'd see their habitation's floor plan: Chamber Nine was a D-shaped cavern sculpted by millennia of eddying water, its floor slanting sharply upward toward the cave wall. A flatter area about the size of a bathroom served as their living and sleeping quarters. It was nearly impossible to get comfortable. The realization that they were stuck here brought on more whimpers of fear and longing. When one boy started to cry, the others would hold him and try to cheer him up. All of them had wept at one point or another. The cave leaked moisture everywhere, walls were damp, and while it provided them with drinking water, the sogginess was maddening.

Four days in they began raving with hunger. The boys were skinny to begin with—soccer players who would shed calories by the hundreds chasing that checkered ball around the pitch in the sauna that is northern Thailand's jungle-carpeted foothills. But now their bodies had pillaged their stores of glucose and turned to the only other source of calories available, the fat and muscle clinging to their bones. The cave offered nothing to consume—if they had been lost in the jungle above they could have lived for weeks on the wild bananas, breadfruit, lychees, and pineapples that grow everywhere. But here there was only mud and rock and the occasional cockroach-size translucent crab skittering across the mud.

Not only had Night missed his birthday, but his cousin Nick's birthday had come and gone on Sunday, June 24, their second day in the cave. Like Night, it was Nick's first season with the team; he'd only joined a few weeks before, playing alongside Night on the 16-under team. Nick had turned fifteen, but as with Night's missed birthday, the boys didn't celebrate—no hugs, no cheers. It's not that they didn't care, it was just that they were too depressed and scared to celebrate anything.

Since arriving, their bodies had not just been cannibalizing muscle and fat, but also the oxygen in the chamber. Their lungs would suck in air, scrub out about 4 percent of its oxygen, and expel the leftovers back into the womb of the cave. Slowly but surely, the amount of oxygen in the little den they inhabited a mile and a half into the mountain dwindled. Breathing was becoming more labor intensive, and combined with the hunger—for food, for their beds, for their parents— the thin gruel of air left them weak.

In an effort to distract them from their hunger and fear, Coach Ek would lead them in meditation, focusing on the Buddhist tenet that "there is no body to be born, no body to die," and therefore nothing to fear. The former monk also felt he had to give them a sense of mission, so he'd set them to work clawing at the cave. Toward the back, in a corner, was another little side cavern. It's where the young athletes, inculcated in perseverance through years of soccer training, began trying to dig their way out. Slamming rocks into the wall, they scratched at their prison like chickens until their fingernails cracked.

Chapter Six
The Foreigners

Even just a hundred yards into the cave it was perpetual night, and that's where Ben Reymenants found himself, staring at what he called a "cappuccino whirlpool." He was in Chamber Three, which had now become the base of diving operations.

A cocky Belgian with an impish smile, Reymenants co-owned a dive shop in the resort town of Phuket that offered courses in cave diving, and when his friend Ruengrit had called to say they needed cave divers he offered his help. He'd already had all his dive gear packed because he and his wife were planning to leave the next day for a vacation in the Philippines. He put that trip on hold, asked his wife to stay put, and hopped the two-hour flight up north, arriving late on Tuesday, June 26.

Now about a mile into the cave, late on Wednesday he found himself at the water's roiling edge with a few Thai SEALs, peering into froth. The water and the close calls had apparently worn them out, so Reymenants and two buddies, Ruengrit and Bruce Konefe, a fifty-seven-year-old former American marine who now writes cave-diving manuals, debated diving in. Ruengrit and Konefe quickly realized that this was beyond their respective abilities. Ruengrit didn't feel comfortable and Konefe was too big to fit through some of the squeezes; Reymenants thought he could do it, so he jumped in.

The water was ferociously cold and the current was so powerful it had been ripping off divers' masks. Kicking as hard as he could, pulling himself forward rock by rock, he progressed a measly 150 yards. He was trying to lay line—unspooling rope and setting down weights or tying it off in key locations to point future divers in the right direction. The bag of climbing rope the Thai SEALs had given him weighed about twenty pounds, forcing him to swim even harder. While this might not seem extraordinarily heavy given the dozens of pounds of gear a diver normally carries, as one diver told me, "these bags were bulky and floaty and caused a huge amount of drag . . . and totally appropriate given the conditions." Reymenants couldn't believe it. As he later recounted, "I was as exhausted as if I'd climbed Mount Everest."

One of the trickiest sections of the cave, with the tightest restrictions, occurred along the stretch between Chamber Three and Chamber Four, where the rock ceiling dropped to barely two feet off the bottom. Reymenants says he had to let air out of his buoyancy control device (BCD)—or inflatable vest—to allow him to descend to the gravel on the tunnel floor. Alone in the water, he says he felt a suction flow—a diving term for a directional change of the current. It was pulling him where he didn't want to go, "an absolute no-no in diving." The man who had spent days in hyperbaric chambers, who had plumbed countless caves, who taught cave diving, was now in exactly the situation he dreaded.

Suddenly, he says, a tractor beam of current started yanking him toward the blackness.

"You get a red light going on in your head, and then a second red light, then a third, and a general alarm blares," he says. "It was my first day, I was alone, and I chickened out."

It's what likely happened to many of the Thai SEAL divers as well—who may also have realized how close they came to death. Getting inhaled by the cave's currents to a blocked passage could lead to disorientation, wherein a diver could burn through an entire tank of air poking around for a way out and asphyxiate. Or the diver could get trapped, which could portend a less mobile route to asphyxiation. Unknown chambers present the challenge of unknown or unseen hazards that could sever air tubes or pluck a regulator out of a diver's mouth. Sometimes that leads to panic, which, if uncontrolled, sends a diver into a spiral of progressively worse decisions and ultimately to death.

The swim back was easier. The current basically spat Reymenants out into Chamber Three. Collecting himself and his gear, he passed the Thai SEALs and stopped to talk to Ruengrit.

"We should tell the [SEAL] commander that we cannot do it," Reymenants said to his friend. "It's suicidal." He didn't let on, but he was shaken. Hours later, in a sweaty slumber, Reymenants had nightmares about it.

On his way out, somewhere between Chambers Two and Three, Reymenants met the British divers Rick Stanton and John Vollanthen. The two groups formed an instant dislike of each other—tension that would persist throughout the entire rescue attempt.

It was after 10 P.M. on Wednesday, June 27. Stanton and Vollanthen had been on the go, packing and travel-

ing and unpacking, for about thirty-six hours. It was
raining when Stanton and Vollanthen were driven to
the cave complex. Over the noise of raindrops pounding the windshield and the metronomic clicking of the
wipers, Vollanthen turned to Stanton and said, "I hope
you're not planning on wearing your inner tube when
we get there." Most divers, like Reymenants, use some
form of inflatable vest or "wing"—which retail at over
$300 and control their buoyancy in the water. But Stan-
ton favored a fourteen-year-old patched-up car inner
tube that he'd jury-rigged for just that purpose. "But
it's my lucky wing [inner tube]," responded Stanton,
and reminded Vollanthen that he'd used it in all of their
previous successful rescues.

Vollanthen genuinely dreaded accompanying Stanton with that amateurish-looking donut on his back as
he walked into the cave before hundreds of cameras.
"You'll look like a cockwomble* before the world's
press," Vollanthen warned him—and like a mischie-
vous spouse, Stanton may have relished inflicting a bit
of good-natured embarrassment upon his friend. And
Stanton was indeed photographed later in the search
trudging through the mud towards the cave wearing
that black inner tube on his back, prompting a flurry of

---

* "Cockwomble" is a Britishism for a fool.

questions on Facebook—the very kind Vollanthen had anticipated—asking if Stanton was learning to swim.

When they finally pulled into the cave site, there was so much traffic in the central parking area outside that their van had to stop well short of headquarters. It made them easy prey for news crews hungry for anything new. And they swarmed, dozens of cameras blitzing them with light. Microphones were shoved in their faces; people were even live-streaming the "event" and taking selfies.

"What the fuck is going on here?" Stanton muttered to Vollanthen. "This is total chaos."

The ex-firefighter did not particularly like the media and he liked disorganization even less. The pickup that was carrying their gear was stuck in the human traffic jam and could go no farther.

The media certainly knew they were there, but the British team that included Rob Harper couldn't find an official to register them. There were dozens of military police in uniforms and soldiers, but no one seemed quite sure what to do with the suddenly famous Brits. And the Brits had no idea where to go, either.

"It felt like we'd been abandoned there," said Stanton. But Vern and his partner Tik were there, and the couple instructed the divers—whom they had never met before—to grab their gear and follow them to the ranger station. There they commandeered a room, dumped their gear, and began to assemble their diving kit. Outside they found one of the Thai SEALs' compressors and filled up their tanks.

Regardless of the commotion, they needed to see the inside of the cave. After an hour or so Stanton, Vollanthen, Vern, and Harper—the group's main liaison with the British Caving Rescue Council, arguably the world's preeminent caving rescue group—headed in.

In the practice of caving, the compulsion to discover subterranean patches untrammeled by man eventually slams into the inevitability of submerged passages. If caves breathe, as cavers like to say, then mountains drink—absorbing huge quantities of rainfall, with caves acting as their veins, flowing with runoff. Naturally, some early cavers evolved into cave divers. Caving was already dangerous. Early cave diving was even more so—so dangerous that divers sometimes use the analogy that an early cave diver had the life expectancy of a World War I pilot—somewhere between eighteen and forty hours.

Some cavers in trouble didn't die, but simply became stranded. Stranded cavers needed rescuers. In the United Kingdom, little groups of rescuers eventually banded into the British Cave Rescue Council, which became the main British cave-diving rescuer umbrella organization. Its vice chair, Bill Whitehouse, says adrenaline is the enemy of divers. Many have a Zen-like quality—at least in the water. And perhaps none are as near flatline as Stanton and Vollanthen. Combined, they had been cave diving for nearly sixty years.

Rick Stanton grew up in Essex, England, just outside London. When he was thirteen, he sat down with much of the rest of the country to watch the documentary *The Underground Eiger*, about two British explorers' record-breaking effort to crack the West Kingsdale Master Cave, not far from Vern's hometown in Lancaster. The film features a pair of shaggy-haired divers, Geoff Yeadon and Oliver "Bear" Statham, as they spent years hunting for the ultimate cave. They eventually managed to connect two cave systems—about six thousand feet of underwater hell—squeezing through obstacles with names like "Deadman's Handshake" while narrowly avoiding rock falls, asphyxiation, and drowning and using what would now be considered Stone Age diving equipment. It deeply appealed to the young Stanton, who, like so many other British diving greats, began his caving career with the local Boy Scout troop.

Through high school and college he continued to cave and picked up diving—much of it self-taught. He began joining ever-more-difficult expeditions, even after he joined the British Fire Service in 1990. Vollanthen also became a caver through the Boy Scouts in the late 1980s. He studied electronics at university and now runs his own firm, employing about ten people. Vollanthen has had one set of experiences that sets him apart from Stanton—he's a father to a thirteen-year-old son. When his son was old enough, he enrolled him in the scouts—thereby hoping to get him interested in caving as well. Though one of the world's most experienced cave divers, Vollanthen got certified as a caving instructor so that he could lead his son's scout troop into caves. In a short time, he went from having almost no experience with children to becoming something of an expert on tween boys. This helped him develop a bedside manner that would prove enormously helpful later on.

Over the years Stanton and Vollanthen would become known as the United Kingdom's top cave divers. The pair had worked together for two decades; along with another self-taught British cave diver, Jason Mallinson, they had set the world record for longest exploration dive of a cave—the Pozo Azul system in Spain. Their expedition lasted two and a half days and took them 28,871 feet—just sixty feet shy of Mount Everest's height.

Years earlier, in March 2004, Stanton and Mallinson were called to the Alpazat cave complex southeast of Mexico City. Technically, at least, it was a straightforward job. A group of British soldiers had planned on spending thirty-six hours exploring the cave. When a flash flood cut off their exit, they got stuck. Six soldiers clung to a ledge above a subterranean river in relative comfort: they made hot meals, even slept in sleeping bags. In a series of six forty-five-minute swims Stanton and Mallinson pulled the troops to safety, one at a time. The water was relatively clear and warm.

The rescue itself took a day. The diplomatic fallout would linger for over a decade.

The purpose of the mission that brought the soldiers to Mexico was unclear, perhaps shadowy, and the British officer in command—who had escaped the cave before the flash flood—reportedly waited for days before alerting local officials. Once he did, he allegedly refused to work with them, demanding that only a British team conduct the rescue. This may have had something to do with the soldiers having entered Mexico on tourist visas, prompting then president Vicente Fox to rifle off a letter to London: "We are asking the British government to tell us whether these people are military personnel, and if they are, what they are doing there."* Regardless of the politics, Stanton and Mallinson completed the rescue successfully, and their renown as the world's preeminent rescuers only grew.

Based on that reputation, Stanton and Vollanthen would also be called up for rescue dives in Ireland, Norway, and France. In the last of these, in October 2010, the French government made a special request for Stanton and Vollanthen to rescue Éric Establie, forty-five, who was mapping the Ardèche Gorges underground tunnel complex in southern France. Establie was a world-class cave explorer, the owner of an underwater engineering business, and a friend of both Stanton and Vollanthen. They spent eight days prowling the sumps about two thousand feet into the cave, finally pushing through a rock slide to find his body. Both said they had expected to find him alive—hoping that somehow he'd found an air pocket and clung there. It was a painful blow to Stanton and Vollanthen, who left his body in the cave because they were unable to pull it through the large rock fall that had effectively sealed off his exit and killed him.

---

* The most likely answer to Fox's question is that the soldiers, from various branches of the British military, wanted to go caving and maybe do a little training, but wanted to avoid the red tape of going through official channels.

Only a small subset of the earth's population is comprised of cavers, a smaller subset of that is comprised of cave divers, and a select few members of that group are cave-diving rescuers. In the insular world of cave diving Stanton and Vollanthen had shown time and time again that they understood how to navigate some of the most complicated underwater passages on earth. They were the specialized rescuers who could bring hope to seemingly impossible situations, which is why they were called to Thailand.

Hoping to get the lay of the land, Rick Stanton, John Vollanthen, Vern Unsworth, and Rob Harper zigzagged around the hundreds of workers manning pumps and hauling in cables and snarls of electrical wire. Though well over a dozen pumps were now draining an output measured in the hundreds of thousands of gallons a day, they were unable to compete with Mother Nature's input of many millions of gallons a day. The Brits struggled to comprehend the chaos. At one point that night a translator attached himself to the team. Stanton asked him which group he was with. The man answered, "No group—I just came to help." He had no problems getting into the cave, and Stanton said he'd proved himself extraordinarily helpful, but Stanton could not help shaking his head at this lack of coordination. He worried that it would lead to somebody getting lost, hurt, or killed. Alongside their new sidekick, it took the Brits over forty-five minutes to reach the passage leading from Chamber Two to Chamber Three. It was near midnight between Wednesday and Thursday when they heard the sound. It was the same noise the boys had heard days earlier.

"It was deafening," recalls Vern, "the floodwater was coming in at such a pace!"

The two divers and Vern hung back as Harper waded through the sump from Chamber Two to Chamber Three. In the few minutes Rob was gone, the water rose relentlessly. Vern started shouting down the hole, "Rob, get back!"

The sixty-five-year-old Harper is bespectacled, round, and topped with a jowly face that belies his agility as a caver and his fearlessness. But bravery wasn't what kept him in place. He couldn't hear Vern and the other men shouting for him to get back. He was so far into an alcove in Chamber Three that he didn't notice the narrow passageway filling up behind him. He had no idea that the other three men were watching as the rising water sealed him in, one inch at a time. Harper had no diving gear, and getting stuck would have forced a rescue.

"Get the fuck out of there!" came the more urgent shouts, as the water neared the roof of the sump.

Harper still couldn't quite make out what was being said.

"What's happening?" he yelled back. Scanning around, he immediately noticed the sickening surge.

There was less than a foot of air between the water and the jagged roof of the cave. By the time Harper wiggled back, he had just two or three inches of breathing room. Tilting his head upward, he bobbed his way through, sucking in available air in the thumb's space between his lips and the ceiling. In his dash out he'd dropped his backpack, which was actually Stanton's—Harper had been carrying it for him. So Stanton threw on his diving gear and chased his bag. It was not the start the preeminent rescue group had hoped for.

Within two hours or so, the water had begun filling the second chamber and pooling in the first. It was now clearly visible from the mouth of the cave. It didn't help that some of the pumps were not working. The operation had switched from diesel pumps to giant industrial ones that ran on electricity, but some lacked parts and others demanded quantities of electricity that could not be wired all the way into the cave where they were needed. And because electrical current weakens over distance, some of the pumps deep inside the cave that were wired to generators hummed at about half power. Given that hundreds of rescuers were standing or swimming in water, and that tangles of electrical wires now wound through the cave, there was a persistent threat that some poor soul would be electrocuted.

On the fourth day of the search, it happened. A soldier accompanying a pump crew staging equipment deep in toward Chamber Three began wading into a canal when, *zing*, he was electrocuted. He suffered minor injuries but survived. A message was passed back to the cave's entrance on the little Israeli walkie-talkies, and within minutes the message bounced back inside: nobody move. No one was allowed to even set foot in a puddle until a team of electricians inspected the length of the cave. It was like a game of freeze involving hundreds of workers. It took hours for the electricians to test fuse boxes and cable connections. Finally, the all clear sounded, and the cacophonous slap of footfalls in mud, the grunting of workers, and the whir of pumps started up again.

The news of the rising water was relayed to the Thai military, which suspended diving operations and ordered a hasty evacuation of the entirety of the cave beyond the entrance hall. Hundreds of rescuers, soldiers, volunteers, electricians, and pump operators straggled out into the steamy jungle. Among them was an American Special Forces team, the U.S. Air Force's

353rd Special Operations Group, headed by a tall major with a head shaved smooth named Charles Hodges.

Early on the morning of Wednesday, June 27, U.S. Indo-Pacific Command, overseeing U.S. military operations for 52 percent of the globe, from the Indian subcontinent nearly all the way to Hawaii, received an urgent request for assistance from the Thai government. The Thai Special Forces needed all the help they could get. The assignment that morning filtered down to Major Hodges during his workout at the air force's county-size base in Okinawa, Japan. Over the next few hours he assembled a team, packed multiple pallets' worth of gear and enough rations to last a few days, and loaded it all onto an Air Force C-130—a rugged pig-shaped transport plane. And then they waited, team belted in, engines running. One of the team's para-rescuemen, a cave-diving hobbyist and staff sergeant named James Brisbin, was making his way back from vacation. He raced from Okinawa's commercial airport to the air force base and scampered onto the plane, still wearing his board shorts and flip-flops.

The members of Hodges's unit are trained to be human Swiss Army knives, capable of dealing with anything that comes their way. They are trained to parachute into hostile territory to treat downed military personnel, call in air strikes behind enemy lines, and direct air traffic in disaster zones. But they are also mountain climbers, scuba divers, sky divers, geologists, surveyors, and organizational whizzes. The unit provided Haiti's only air-traffic control in the hours after the country's 2010 earthquake, surveyed runways during the 2011 Fukushima tsunami and nuclear disaster, and led some humanitarian efforts following the 2004 Indonesian tsunami. Chaotic, fluid emergencies are their preferred element. Hodges describes their job even more succinctly: "We solve problems." In Thailand that meant "figuring out what was going on. Figuring out which resources are available to us. Connect the right people together. And then effect a positive outcome."

When they landed at Chiang Rai airport, pulling up beside the Thai Navy SEAL planes, they were greeted by an embassy liaison who had scrounged up vans and trucks to take them wherever they needed to go. Hodges sent most of the team to sleep, but he personally needed to survey the cave to see what they were up against. He selected a small team. They arrived at the cave just past 1 A.M. on Thursday, June 28, not long after the British team. It had grown relatively quiet. The first thing Hodges noticed were the mud stains about twenty feet up in the first chamber—indicating how high the flows can get. But as they went deeper into the cave that night, like the rest of the rescuers, they didn't immediately notice the water rising. And initially they ignored calls to pull back.

"A few minutes later," Hodges recalled, "we heard more urgently, 'Get back. The mouth of the cave is starting to fill with water now.' Even though where we were, we really didn't see much. And so we started walking back. It took us two or three minutes to get back to the mouth of the cave. And in that five-minute or so period, water had started coming in. And the floor of the mouth of the cave was already flooding. A fifteen-hundred-square-foot area already had two to three inches of water in it. And so that all happened in about a five-minute period."

Hodges had two realizations as he watched the water eat away the remaining floor space of the cave's entrance. The first was that the water didn't only come from a single source, but seemed to ooze in simultaneously from everywhere at once. The second was that conditions within the cave could change in an instant.

The scattered retreat from the flooding cave that night spurred the Thai Navy SEALs to order the suspension of all diving operations, but they continued their efforts to stockpile scuba tanks. Multiple reports indicate that these suspension orders included the British divers.

But the next day, Brits Stanton and Vollanthen pulled on their Wellington rubber boots to protect their feet, grabbed air tanks and spools of guideline, and headed in for an exploratory dive. They were not sure if the ban applied to them, and frankly, didn't care. They needed to inspect the cave. They were celebrities in camp already, so the guard at the mouth of the cave didn't challenge them—perhaps figuring the eminent divers could be trusted to look after themselves. When they stepped off into the new body of water at the edge of Chamber One, they expected to be gone for seven hours. Vern and Harper decided to wait for them outside the cave.

Just three hours later, lights started flickering through the black water from below. Onlookers saw expanding bubbles, the telltale signs of humans breathing underwater. Moments later a phantasmal figure started to flail toward the surface in Chamber Two. It was a bedraggled Thai man in orange coveralls. Observers were baffled. Stanton and Vollanthen had intended to explore the cave, but when they surfaced into Chamber Three and clambered out of the sump, four miserable-looking men came dashing into the beams of their helmet lights. They were pump operators with a water-management team who'd been marooned beyond the sump during the pell-mell evacuation. They had been stranded in Chamber Three for nearly twenty-four hours, the water and the rock blocking all cell and radio signals out.

The rattled Thai crew had been perched on a rock above the sump and were pointing frantically to the water. Stanton, with his typical understatement, remembers that the water was rising "quite quickly" and the men "were quite keen to get out."

Stanton and Vollanthen had been unprepared for what's called a "snap rescue." They lacked the extra masks, fins, and even unattached tanks. Since experienced cave divers typically carry at least two tanks, each connected to a regulator—the mouthpiece attached to a hose through which the diver breathes—supplying the workers with air during the quick rescue would not be a problem. The problem was that it forced the rescuers to be physically attached to their panicky wards. Letting them go could mean letting them die. They easily cruised through the first part of the submerged cavern, but as they neared the silvery surface, that changed.

The dive through the sump separating Chambers Two and Three was short, barely forty-five feet. An amateur free diver (who didn't fear bashing his face against a rock or two) could have done it. But the workers couldn't have known that, plus they were not "water people," as Stanton put it, and perhaps the safety of a dry rock appealed more than the unknown of the glimmering black pool that was their only exit.

The Brits had to use sign language to explain the mechanics of the rescue. They showed each man the spare regulator that was attached to a second tank on their backs, put it into their mouths, and acted out the swimming. They would be taken out one at a time, each diver with a single worker. The workers seemed to understand. What couldn't be passed on via hand signals, though, was the warning, to just let the rescuers do their job and not to panic.

It seemed to go well until the workers spotted the silvery surface. "Coming up, when they could see the air space and were trying to hurry up to get to it," Stanton told me later, "they were trying to get away from us, and we were physically trying to restrain them. It was like an underwater wrestling match."

Panic is the single greatest enemy of divers, and during this initial, unexpected rescue, the flailing "casualties," as the rescuers call them, clearly panicked. Flailing casualties could rip off face masks, sever regulators from their tanks—and, worse, grab onto their saviors, potentially dragging both down to a watery grave.

Typically, divers prefer bailing out "casualties" who are either capable divers themselves, compliant, or completely inert—meaning unconscious or strapped down. The experience with the pump operators left Stanton and Vollanthen concerned. If it was this hard to pull four adult men who could swim out of a short underwater passage, how were they going to drag twelve scared kids and their coach through an underground river more than one hundred times as long?

Parting ways, the Thai workers thanked them and shuffled off into the gloom. Stanton called the parting "a little anticlimactic." With few people in the cave as the group emerged from the water, no one really noticed or seemed to care that four human lives might have just been spared. One team did make careful note of the British duo's resourcefulness—the Americans, who had only pulled into camp hours earlier. If there had been doubts about the Brits' ability, they were silenced.

In the aftermath of the flooding that had left the cave submerged up to the entrance, the consensus at camp late on Thursday, June 28, was that diving the boys out was not an option. The current was unswimmable, the challenge insurmountable. At the very best, it was a bad option. The Thai Navy SEALs were already beaten up. The divers' hands were being flayed by scrambling through passages carrying anywhere from forty to eighty pounds of gear on their backs.

Their feet were succumbing to rot—the skin softened by water and then shredded by the shark's-tooth rocks, even through diving booties; the cave was a breeding ground for bacteria and microbes, so infection crept into every crevice. The SEALs' Facebook page began asking its 2 million followers for donations of medication and space blankets for the troops freezing from hours-long exposure to seventy-degree water.

Stanton and Vollanthen had been told the conditions were not likely to improve—that typically, once the cave fills, its insides continue to churn until the monsoons wring themselves dry. At best it would be weeks before they got a chance to dive. So the Brits threw in the towel. "We don't want to wait here for weeks, just for confirmation the children are already dead. It makes no sense for us, we have lives to lead," thought Stanton. Also, a dozen dead boys was a grim reality they'd rather not have to face. Added to it all was frustration with the cultural differences and the search and rescue mission's disorganization. So the world's elite cave divers quietly looked into flights home, telling the Americans and their British Embassy liaisons that there was not much anyone could do for the boys.

Chapter Seven
The Shadow Quartet

They felt half dead. But the dead can't smell.

By now the boys had mostly grown used to the stench. Though Coach Ek had the boys dig a latrine pit, after several days, the urea in their urine evaporated, crowding the chamber with the stinging smell of ammonia. It was accompanied by the distinct bouquet of human feces, sweat, and fear.

The boys and their coach could tell by their digital watches that a week had gone by, but they almost couldn't believe it. They didn't know if anyone knew where they were or if anyone was coming. Tee, the captain, would hold his mouth open under a stalactite and swallow drop after drop until his belly felt full. They hadn't eaten a meal since Saturday morning. With no light, their circadian rhythms were off and sleep came fitfully, each one of them squirming on the mudpack to try to nudge a bare shoulder into a more comfortable position. Eleven-year-old Titan's jersey sagged to his lower thighs. He had been pulling it over his knees to keep warm. As their necks thinned and the shirts stretched, their collars started to hang, revealing sharp collarbones. The muscled trunks of their legs were turning twiggy.

After about forty-eight hours without food, the human body begins ketosis. When the nervous system is unable to locate energy-rich glucose to consume, it orders cells to consume muscle and fat—and one of the byproducts of that chemical process in the body is increasingly pungent urine. The bigger boys, including Tee, Night, Nick, and Adul, had more body mass stored. But little Titan, who weighed less than seventy pounds, suffered fainting spells. Biw, with his fleshy face, was a class clown—often earning frowns from his English teacher, Carl Henderson, for doodling or scarfing down snacks during class. They tried not to think about food, but then that would only trigger more thoughts about food: basil fried rice, crispy pork, fried chicken. And then sometime on the fifth day, the ketosis fully kicked in and the hunger pains gave way to a great weakness, sometimes accompanied by what felt like flu symptoms. Now the only thing in their mouths was a strong metallic taste that made their breath reek. They slept most of the day, hugging themselves to keep warm.

The temperature in the cave was 73 degrees, which would seem comfortable, even ideal. But this natural air conditioner was now a curse. Their bodies' self-cannibalization had put their internal thermostats on the fritz. They were freezing—all the time. To keep warm they had cleared out an area where they huddled together during those semiconscious sleeping hours. Sometimes they slept by the water, desperate to hear the clatter of rescuers or their names echoing through the cave. Sometimes they heard sounds they could not explain, like dogs barking or children playing.

In their waking hours, Coach Ek instituted a strict flashlight protocol, allowing the boys to use only one light at a time. Their sense of time became soupy. But in that haze a few days in, their ears pricked up. The coach shushed the boys. They listened. Was it divers? No. The sound of rushing water became unmistakable. They pointed their flashlight to the stream below and watched the water course past. Within minutes they noticed it wasn't just surging forward, it was also push-ing upward. They backed up the slope. And the water kept coming. The coach watched it rise nearly ten feet and ordered the boys to huddle higher up. They had been waiting for rescue, but suddenly they realized they might have to rescue themselves.

Their routine was basic: guzzle water from the rising stream until their bellies felt full, then dig. Pee and repeat. They hacked at the brittle limestone for hours at a time, chipping away at their mountain prison. If the coach thought it was futile, he didn't let on. Help had not arrived, and it was vital to keep the boys busy and focused on a goal day after day. They had carved out a mini-cavern several feet into the wall. After their workdays, they would creep back down to water's edge to drink (if they found a clear current) and wait for the sound of rescuers.

When they talked, it was no longer about food, but fantasies of how they would escape. With each fluctuation of the water level they would discuss whether the trend would continue, and whether now might be their chance. "Maybe now," they would say. Maybe not. Now? No . . .

They would test the water in their tomb. They realized its depth was only about an arm's length above Adul's head. He was brave enough to sink under—letting his toes touch bottom. But on either side of the passage, the ceiling dipped down to the water. Many of the boys went in, but no one was willing to dive into those sumps. Coach Ek wouldn't let them—besides, by now no one had the strength. They would keep tabs on who had tested the water, who was due to go in, and who wouldn't.

Between the now-haphazard digging and talk of their confinement, there was little else to do. As the days dragged on, their weakness kept them confined mostly to their perch.

As the boys clawed away from below, the Thai Third Army was hammering away from above—literally, with hammer and chisel. Dozens of men from the Third Army's Thirty-seventh Military District in Chiang Rai had stuffed themselves into the crevasses thousands of feet above the boys, pecking away at the mountain, hoping it would reveal a mysterious entranceway. In fact, by Thursday June 28, when merely diving to Chamber Three required herculean effort, the rescue operation zeroed in on finding alternative ways into the cave: by drilling a rescue shaft that would somehow locate the boys, or finding an as-yet-unrevealed alternative cave entrance which would bypass some of the worst of the flooded tunnels and therefore give divers a fighting chance. And the other option relied on reducing the water levels in the cave—pumps inside the cave were barely maintaining the status quo, and proved finicky.

The entire country had mustered to help the boys.

Dozens of multimillion-dollar drilling rigs were parked by the side of Mae Sai's roads. Their operators, sitting in their cabs smoking cigarettes, waited for orders. Owners of industrial pumps had chugged north to Mae Sai, hoping to help. Scuba gear was coming in from all over the world. Even Thailand's king, Maha Vaji-ralongkorn, sent kitchen staff and trucks to feed the workers. He would later also donate diving supplies.

In a matter of days, thousands of volunteers had arrived in Mae Sai. What had been the cave's base camp turned into a jumble of mess tents and medics' stations. The stairs leading up to the cave from the parking lot below had disappeared under orange and blue rows of thick hoses so densely packed that together they looked like a mud-slicked Slip 'N Slide down the stairs. Only a narrow section of stair was left for workers to trudge up.

Those hoses led to a series of pumps arrayed inside the cave. The squat cylinders about the size and weight of a washing machine needed constant maintenance, and though the pumps themselves were enormously high-tech, getting them into the cave utilized prehis-toric technology. Thai soldiers would thread a thick green bamboo pole beneath a barrel-shaped pump's steel handle—which resembled a bucket handle—and hoist the pole onto their shoulders. To move a single pump required eight men. The hoses themselves weighed hundreds of pounds and required additional platoons of soldiers to hoof them in on their shoulders. With those orange, red, and blue hoses, the floor of Chamber Two looked like a messy fuse box. Physically wrangling and maintaining hoses that were being tramped on by hundreds of people a day kept dozens of workers busy.

There had been no evidence that the boys were still alive. Governor Narongsak and other officials had informed the press of the plan to try and drill a relief well to the boys in the cave, but no real headway had been made.

A man named Thanet Natisri had pulled into camp at about 7 P.M. on Thursday, June 28, and was quickly ushered into the cave for a briefing. He had come prepared with maps his associates at the Geospacial Engineering and Innovation Center in Bangkok had provided—state-of-the-art maps detailing the contours of the mountain and the estimated elevation and depth of the area above the Pattaya Beach section of the cave. He presented this information to the search mission's chief engineer and a few military commanders, who were impressed because they hadn't had access to that information. At 9 P.M. Thanet, who had been running multiple water-management projects in

Thailand on and off for the past five years, started by lobbying Governor Narongsak for a seismic scanner to peer inside the mountain. The scanner, which required a permit from the Thai Department of Mineral Resources, worked by inserting firecracker-size charges a few yards into the earth, setting them off, and measuring the reverberations. Similar to sonar for mining, it's often used to locate oil by detecting areas with different density—like cave chambers.

Thanet had hoped that seismic tests could help detect the cave chambers. They could then drill down into them one by one, perhaps lowering microphones to listen for the boys. His plan called for dismantling the smaller drills at rescuers' disposal, then slinging them to powerful Russian Mi-17 helicopters and reassembling them once they landed on the mountain. They had the manpower, they just needed a piece of paper permitting the seismic tests.

But that Thursday, June 28, Narongsak, who has multiple master's degrees in engineering, overruled both his engineers and Thanet—denying the request for a seismic-scanner permit. He approved of continued attempts to drill down to the boys but not the seismic tests, concerned that the boys might somehow be harmed thousands of feet below or that some seismic reaction might block access to the boys, complicating a future extraction. After all, they had no idea whether the boys were trapped only by the water, or by a landslide, or by both. It was also Narongsak's responsibility to ensure the safety of the thousands of soldiers and volunteers now there. Instead, weaker sensors that could only penetrate a few hundred feet into the mountain were approved. But Thanet argued that drilling down into the mountain without a precise target in hopes of finding the boys could take years—or might be entirely impossible. Still, he was overruled. No seismic testing.

It would have been easy to dismiss Thanet. A boyish thirty-two-year-old who looks twenty-two, he's gangly, with a flop of thick hair dyed slightly orange. When he's not managing water for rural communities in Thailand, he runs a Thai restaurant in Marion, Illinois. His was perhaps not the most illustrious of résumés at the cave site, given the number of generals, professional politicians, and world-renowned rescuers present. Yet he had impressed local government officials and military officers a month earlier in a nearby province during a lecture on water-management techniques. His Thailand Groundwater Recharge Project had helped farmers manage aquifers in the dry season and divert the increasingly severe floods in the wet.

And that's why his newfound military contacts had called him: he seemed to be a person with the magical power to make water appear or disappear. Thanet immediately got to work as the search-and-rescue operation's lead outside consultant on water management. If water was the enemy, Thanet knew he couldn't defeat it. But he could divert it.

One of the extensions to the Tham Luang cave that Vern had explored was likely the outlet for a stream that ran out of the cave, draining a few hundred yards south. It was called the Tham Sai Tong. That part of the cave was at a lower elevation than the main entrance, about six hundred yards south, and Thanet concocted a plan.

He rounded up thirty drilling rigs. Then he drove around Mae Sai hunting for the men smoking cigarettes outside their industrial pumps. Bored out of their minds and eager to volunteer, they were easy to recruit. He then collared three hundred soldiers from the Thirty-seventh Military District lent to him by Colonel Losuya. Then, without seeking permission, he sent them down the muddy track to Tham Sai Tong. Excavators felled trees, leveled what had been one of those lovely little picnic areas near the cave, and carved out a wide flat of grayish mud. Then began a massive industrial drilling and pumping operation, in the middle of the jungle, six hundred yards from the camp. And no one knew about it.

The goal was also audacious: to lower the entire water table of an area nearly the size of Manhattan. Governor Narongsak had already issued orders to reduce the water table near the cave entrance, but space was limited for the drills and pumps and the generators they required.

In theory, Thanet's idea was to create an artificial drainage point for the cave—on the assumption that if the water had a place to drain, it wouldn't back up as much. The plan looked similar to the method used in drilling for oil, only Thanet was seeking to extract water. Using the drilling rigs, they would drill down until they hit water and then use industrial pumps to extract it.

The mountain was so waterlogged that as soon as the drill bits pierced the surface water geysered ten feet into the air. The pump-truck operators rolled up their pants and worked barefoot, siphoning off and diverting hundreds of thousands of gallons an hour, says Thanet. Later it would be over 2 million gallons an hour. Thanet supervised, sloshing around with a MacBook under one arm and a cellphone in hand, skinny legs sticking out of black rubber mud boots. Sometimes he wore pink knee-high compression socks to discourage chafing and foot rot.

It was only once the project was under way, twenty-four hours after it started, that Thanet informed his superiors. "Retroactively," he said with a wink, "they approved the plan." Whether lowering the regional water table actually reduced the amount of water in the cave is debatable. But what became clear was Thanet's industriousness. He started carrying so many maps, surveys, and lists of coordinates that his friend Colonel Losuya appointed two of his men to be Thanet's full-time aides-de-camp.

Thanet's drilling efforts had opened up a new front in the battle against the water. Between his work and the efforts going on inside the cave, where pumps had been working nonstop to lower the water and reduce the current to diveable levels, everyone hoped that the water would begin to drop.

Thanet next sought out the Americans in the camp—Major Hodges's U.S. Air Force 353rd Special Operations Group, the air force's commandos he had heard so much about. They'd arrived to some fanfare early on June 28 because it was immediately reported that they'd brought in some American-engineered superweapon. It was said to be an infrared device or a satellite that could peer deep into the mountain—perhaps the same one used to locate Osama bin Laden. In announcing their arrival, Thailand's deputy prime minister, Prawit Wongsuwan, told reporters: "The United

States Pacific Command (USPACOM) has sent thirty staff with equipment to help penetrate the cave walls."

A parade of Thai military commanders began stopping by the Americans' little camp eagerly asking about their mountain-penetrating device. There was a hitch, though: no such system existed. Yet the truth did not seem to stifle rumors that not only did the Americans know how to find the boys, they already knew where they were.

Which was why Thanet stopped by. He thought perhaps the bald-headed commander, Major Hodges, and his top noncommissioned officer, Master Sergeant Derek Anderson, might be able to help him find a way to harness all those idle drills and get down to the boys. He was informed that the concept of the American superweapon was risible. (Though their kit did include several pieces of equipment that would prove critical later on.) Once Hodges gently informed all comers that really, honestly, they didn't have that equipment—that it didn't exist—the Thai military commanders seemed to lose interest in the Americans. Still, that particular rumor was hard to kill.

Thanet did not find the sensors he wanted, but he did find allies. Hodges, a little older than many majors because he had left the military for the civilian world for a few years—and perhaps wiser for it—was willing to listen to any idea. Anderson, his top NCO, was the same age as Thanet. They came from utterly different worlds but shared a sense of industriousness and an allergy to bullshit.

So a shadow planning operation began to form in the corner of the cave's base camp, creating a four-way nexus between Vern, who knew the cave best; the British divers, with the skill sets to penetrate the cave and bring the boys out—dead or alive; a Thai engineer who hoped to defy the deities by robbing them of their watery lifeblood; and the American soldiers who had the organizational skills to pull it all together.

Which was why the weak central command and lack of organization was so frustrating for them. It took the USAF Special Tactics team days just to determine who was in charge from the Thai army, navy, and civilian side, and what everyone's assignment was.

"We were just trying," said Hodges, "to basically bring everybody into the same room and say, 'Hey, we know what the problem is. Now, let's look at what possible solutions we could all provide and utilize and the assets that we have here, the manpower, the equipment. Our goal is to get all these kids out. Let's try to approach this from, you know, several different angles.'"

Using something the U.S. military calls Military

Decision-Making Process, the Americans began to rank the options for saving the boys from the lowest risk to the highest. Everyone agreed that diving the boys out had the highest probability of failure. They printed out organizational charts and maps, and began to try to centralize information. There were now thousands of people crawling over and inside the mountain, and it would take days more to finally establish clear divisions of labor and coordination.

The process was so excruciatingly slow that the Americans initially expected to send their own teammates in to fish the boys out, though they knew that—like the Thai SEALs—their divers lacked the training and equipment. While they knew the British divers had already rescued those pump workers, it took them a few days more to comprehend the totality of Stanton's and Vollanthen's skill levels and confidence—and to assess their willingness to lead a mission the Brits had made known to be as futile as it was dangerous.

By Sunday, July 1, Thanet's pumping operations had started siphoning a large pond's worth of water per day out of the ground south of the camp. Thanet says that his ragtag army of roustabouts had lowered the local water table by twenty-four feet. Eventually that water would fill a 650-acre lake—nearly the size of New York City's Central Park.

At the same time, the one force that had started this mess also started to comply. Heavy rains still swept through, but they were short-lived. Imperceptibly at first, the ten-foot-tall yellow water gauges in Chambers Three through One started to reveal more numbers as the river in the cave started to drop. Whether it was from Thanet's drilling operation, the pumping that had been going on inside the cave for days, or the decreasing rain was never clear.

In the end, though, it didn't matter why the water was going down—all that mattered was whether the divers would be able to get back into the cave.

Chapter Eight
Jungle Bash

For the past couple of days military and police div-ers from around the world had been assembling in Mae Sai. It was like a commando convention. Thai Navy SEALs and the U.S. Air Force Special Tactics team led by Hodges were joined by Australia's premier dive-rescue group and a Chinese team. Each coun-try's team had a small blue tent sprouting out from the main hub of the Thai SEALs operations tent. Like the Thai SEALs, the Americans, the Australians, and the Chinese were trained for open-water diving and most didn't have the side-mounted tanks that would have facilitated technical cave diving.

The Chinese team also lacked some critical language skills. Ruengrit, the GM manager and cave-diving hobbyist, recounted that on Wednesday, June 27, he watched the Chinese suiting up outside the swelling sump in Chamber Three. They had four main divers, a three-person support team, and a female translator. She was in her twenties and dressed to the nines. Ruengrit explained to the men through the translator that they couldn't dive because the other teams would have no way of communicating with them once inside.

That's when the Chinese commander jabbed a finger at his translator. "You go in with us!"

She was horrified. "She said, 'What, you want *me* to go!?'" Ruengrit recalled with a giggle. Saving the translator from misery, or maybe even injury, Ruengrit stepped in and told the Chinese he'd find something else for them to do.

For the next few days, some of the world's elite military search-and-rescue teams were turned into mules. They began hauling hundreds of air tanks—first to the edge of the flooded Chambers One and Two, and later to Chamber Three, which became the dive headquarters. It was backbreaking. The conditions resembled an enclosed World War I trench—including the stench. When the Thai, American, Australian, Chinese, and other troops actually found solid ground to tread on, it was dangerous rock slicked with mud. Much of the slog with the heavy tanks balanced on their shoulders was in water. When the water was knee high, troops' legs disappeared when they planted their feet. Ankles turned, knees buckled, skin rotted. The buoyancy of deeper water offered a respite to battered joints but was slower going. And toward the third chamber, they were neck deep, forced into a quick scuba dive to the other side of the sump between Chambers Two and Three. They alternately froze from hours-long exposure to 70-degree water or sweltered in their neoprene wet suits.

By now this had become one of the largest cave search-and-rescue operations in history. The rescue effort was allotted virtually unlimited resources. Governor Narongsak had ordered thirteen ambulances to stand by twenty-four hours a day on the off chance that somehow the boys would be miraculously found and brought out. An entire floor at the Chiang Rai Regional General Hospital had been cleared to receive them. Rotating teams of doctors, nurses, and mental health specialists were on call around the clock to treat possible trauma, organ failure, and infectious disease in the event that anyone was rescued alive. "If we find them, we want their families to know that we are ready to care for them," one of the rescue commanders told reporters.

And what many of the hundreds of troops and officers couldn't have known, but might implicitly have felt, was that their lungs were methodically scrubbing the cave air of oxygen. Breathlessness became a problem, followed later on by a chronic crud in the lungs that troops and volunteers would later tell me lasted for weeks. Some would be stricken with mysterious fevers and dropped off the diving roster.

Belgian diver Ben Reymenants was also ready to hang up his scuba tank. Having endured what he had considered a near-death experience, he was rattled and freely admitted that the seemingly jet-powered current was beyond his abilities. But when he reported to the Thai SEALs tent to inform them that he thought future dives were unsafe and he was pulling out, the Thai SEAL commander acknowledged his advice and said, "But I'm going to have to keep sending my guys in." It's unclear if he was calling Reymenants' bluff, but it worked. Said the Belgian, "In my mind I saw these nineteen-year-old boys going in with no experience. At least I can go in to stop them from killing themselves." By June 30 the water levels remained high, but the current had slackened and visibility was marginally better. That morning Reymenants and dive buddy Maksim Polejaka, a bearish Ukrainian who had gained French citizenship by joining the French Foreign Legion, stepped off into his "cappuccino whirlpool." They hoped to lay 400 yards of line. The current was still too strong for them to use regular cave-diving guideline, so they opted in its place for thicker climbing rope that would enable future divers to pull themselves, hand over hand, toward the T-junction rather than having to swim through the current.

Together Reymenants and Maksim put down a couple hundred yards of the rope and turned back. Because both they and the SEALs had been using the rope instead of the thinner guideline, it could only be tied off around big underwater landmarks. So instead of a laserlike, down-the-middle line of the type that cave divers are accustomed to, the rope line ping-ponged between sturdy stalactites or any geological version of a dock cleat that wouldn't crumble under the constant yanking of divers on the ropes. But that zigzagging also meant the rope was often slack, as if a road-maintenance crew had drunkenly painted center lines in an undulating wave.

While the tunnel is mostly a straight shot, it has many tricky cul-de-sacs. Reymenants now found himself in one of them. Walls were everywhere. He was wedged in. The more he kicked, the more silt the current carried into his face. Divers' fins are designed to propel a swimmer forward. There is no reverse gear.

"Adrenaline doesn't help. So I closed my eyes and focused on breathing," said Reymenants. From behind he started hearing high-pitched, muffled sounds:

"Mmmmm, mmmmm." It was Maksim trying to talk through his regulator. He'd caught up and was now gently tugging at Reymenants's fins. When a fellow diver does that, there are two possible reactions: the first is to kick against it, indicating you want to be left alone and are capable of handling the situation yourself. The other reaction is going limp, which is what Reymenants did. As Maksim pulled him out, Reymenants realized he'd been sucked about two hundred feet into a false inlet, a dead end. They swam back, found an air pocket where they could stand, and surfaced for a breather.

Maksim checked his gauge and said he was well into his reserve of gas—that last third of a tank that cave divers save for an emergency. The only direction he could dive was out. But Reymenants was on a rebreather, which recycles the oxygen (17 percent) that we exhale and enriches that recycled oxygen by an additional 4 percent or so. Under decent conditions a rebreather gives a diver about eight hours of dive time. He had the air, and now he had a surplus of adrenaline, so as they floated on the surface, Reymenants told Maksim to wait a few minutes in that neck-deep water. He ducked down and started pulling and kicking. The distance he covered is a matter of dispute—of interest only to the tiny and competitive clique of cave-rescue divers. But Reymenants claims that in about twenty minutes he put down another couple hundred yards of line, pushing well past the T-junction to near Pattaya Beach.*

That day, Reymenants says, a Thai Navy SEAL helped him climb out of the sump at Chamber Three. He was older than the rest, with a taut smile, a thatch of black hair, and a runner's build. He was among a group of reserve Thai SEALs who had volunteered for the search and rescue. He now worked at an airport, where he had picked up a little English, and offered to help Reymenants explain to the other SEALs how far he had gotten in laying the guideline. The man was thirty-eight-year-old retired Navy SEAL Saman Gunan, a husband and triathlete with a face that looked as if it were chiseled out of granite—a face that, within days, nearly every person in Thailand would recognize.

Over the previous few days, Vern, like most of the people in camp, had offered himself up as a pack mule,

---

* Later Reymenants would say, "I had no idea I had gotten within several hundred meters of the boys!" Several officials, divers, and others indecently offered the same retort: "Bullshit, he didn't get that far."

hauling in air tanks. His unique expertise, however, was his knowledge of the cave. With the rains slackening, he was willing to try anything—even digging through a haystack for a needle that didn't exist.

He and Rob Harper spent two days in what he called a "jungle bash," hauling themselves up the treacherous slopes with the members of the Thai military who were still trying to find an alternative route down into the Monk's Series. They tramped the jungle floor, clearing tangles of vines to shimmy into holes about the size of a bathroom sink. Towering dipterocarp oaks, with gray trunks that looked like giant elephant legs, blocked out the sun. Their cast-off leaves formed a biotic mash on the forest floor feeding bountiful fungi.

The men were never dry. Vern's shoes split. He and others suffered from jungle rot in their feet. They slept on the jungle floor, some of the troops falling asleep with rocks as their pillows. He and Harper tried to tell the commanders that their men were toiling in vain. The alternative route likely didn't exist. They believed this because their dear friend Martin Ellis had written most of the guidebooks on caving in the region, noting that in all of Southeast Asia there was only a single known alternative entrance to a major cave system, and it was in Laos. Vern tried to explain to the Thais that Tham Luang ran horizontally and that any vertical shaft leading about fifteen hundred feet down to the main cave system would have been pinched shut by the glacial movement of the mountain millions of years ago. And even if by some freakish chance there was such a shaft, it would be so enormous that they couldn't miss it. The problem seemed to be that there was no way to prove it, and so the Thais argued that the absence of evidence of a side shaft was not evidence of its absence. But given its almost limitless resources and drive to find the boys, the Thai military was willing to toil to the point of the exhaustion of both its men and every possibility.

More frustrating than the search itself was the lack of coordination in search efforts. Mario Wild, of Chiang Mai Rock Climbing Adventures, had spent days with another Thai military team scoping out possible holes on the south side of the mountain. The team would divide its section in quadrants and search for possible leads. When they found anything remotely promising they would call in rope and caving specialists like Wild. Wild and his boss, Josh Morris, had nearly four decades of climbing and caving between them, which had endowed them with the ability to sniff out a possible passage within minutes. The ingredients are simple: a cool draft and proximity to a main shaft. Wild felt they could cover much more ground if the officer just listened to the advice of specialists like Vern and him: a fifteen-hundred-foot vertical shaft to the main cave was a geological near impossibility—meaning that a promising shaft would be so big and would blow with such a draft that they would all likely notice it right away. But he said it was his role to explore possible cave entrances, not dispense advice.

A member of the elite Austrian Mountain Rescue team, Wild understood that efficiency in searches is not just nice to have, it's elemental. But soon after returning to camp from his few days stomping around the Sleeping Princess's head, Wild learned that no fewer than four other teams of rescuers had explored the very same holes he had. And they boasted about it on social media. He was stunned by the inefficiency and the waste of resources. There was also an increase in people who seemed to be more focused on the potential for glory involved than the actual rescue. The more attention the story received, and the more Thailand and the world became obsessed with it, the higher the stakes were for individual rescuers. So many people wanted to be the discoverer of the hidden gateway to the boys. This "glory fever" was a malady that made some rescue teams cagey, said Wild: "When I asked people, 'Where were you today?' people were secretive about it." That kind of sentiment was common and quickly became a fatal organizational flaw, preventing and actively discouraging information sharing that in turn wasted possibly tens of thousands of work hours.

But the time Vern spent exploring the mountain above the cave wasn't all wasted. Up on the mountain he grew even closer to Master Sergeant Derek Anderson and the U.S. Air Force Special Tactics team that had been assigned to clear a helicopter landing zone in the mountains to enable choppers hauling drills to land. The better Vern got to know them, the more he was impressed by their organizational skills and energy.

Anderson, who with his clean-cut features and beefy physique looks like a Special Forces poster child, stood out. He had grown up mostly in Ecuador, the child of a missionary who would fly him out to the remote Amazon to work with indigenous tribes. They would head into the jungle with local guides for two or three days, sleeping under the canopy of palmitos. The family lived in Quito for most of his childhood, with intermittent stints back in their native Syracuse, New York. Many of his friends were Ecuadorians, so he was bilingual and embedded in the culture. By the age of thirteen, he and his buddies would ride buses to climb the Andes, scaling twenty-thousand-foot peaks. He'd always known he'd be either in the U.S. Navy SEALs or in some other Special Forces unit. After high school he enlisted in the air force, which gave him more opportunity for the swashbuckling adventures he craved—mountaineering; high-altitude, low-opening sky dives; and rescues. In 2014 he was named the U.S. Air Force's Pararescueman of the Year.

With his organizational skills and keen mind, he quickly rose to the rank of Master Sergeant, responsible for much of the tactical team's planning and organization—skills that were much in demand in the increasingly chaotic search-and-rescue operation.

Stanton and Vollanthen were not there to make friends. But they did need allies. Vern, whom they had only met when they arrived, quickly became one of them.

Vern continued to explore the area above the cave, hiking to a valley one thousand feet above the cave, where he met locals who live in an area called Pah Mee. They steered him to a sinkhole where a stream seems to disappear into the mountain. Their ancestors had been roaming the area for generations. They knew the cave and freely offered Vern and the military (and anyone else who would listen) a terrifying tidbit of information; Vern, who only went caving in the dry season, did not know that by around July 10 the cave was typically completely flooded all the way to the entrance. It would stay that way pretty much until December.

When Vern finally stumbled back into camp on Sunday, July 1, he met Thanet again. He'd been trying to tell him that his water-table drilling project at Tham Tsai Tong was a waste of time. Now he had something concrete for Thanet. He described that mysterious creek and wondered whether it was feeding the Monk's Series.

Vern told Thanet that the Brits had finally reached the T-junction earlier that day, adding that Stanton and Vollanthen remained concerned about the water flow from the mountain above; they had also noted that the water flowing from the Monk's Series on the right side was moving more quickly. They had offered another key clue: the water from the Monk's Series was both clearer and warmer than the water coming from deep within the cave on the left.

Both Vern and Thanet knew what that meant. The cave acts as a giant icebox. The longer water stays in there, the cooler it gets. The farther it flows inside the cave, the more silt it picks up and the murkier it gets. Clearer, warmer water meant whatever was flowing into the Monk's Series had recently been outside the cave.

Vern urged Thanet to find the source of whatever was dumping all that water into the Monk's Series. It was a simple equation: tackling the Monk's Series would result in less water flowing in (especially when it rained), which would give divers a chance to finally penetrate deeper into the cave.

Thanet and his crew were allotted two military trucks, and they bounced along a dirt track about the width of a doorway, up hills with the vertical grade of black diamond ski slopes—not far from where the American team had cleared that helicopter landing zone. Tree branches thwapped against the truck. Its gears ground and slipped. They hiked the rest of the way up to the bowl-shaped valley called Pah Mee, whose local farmers pointed them toward a stream which Thanet matched with satellite imagery of the mountain. It was the one Vern had described.

This particular stream was a series of waterfalls, tumbling down the mountain and sluicing around three-ton boulders. Thanet had forty Thai troops from the Thirty-seventh Military District with him and a handful of local engineers. Squads of soldiers in knee-deep water dredged up sand from one of the creek's pools; the villagers helped where they could, shoring up the dams. Filling thousands of sandbags, the troops began to pack them against one of the stream's smaller cascades. But they had to drain it somewhere. Industrial pumps couldn't make it up the muddy tracks, so they hooked six-inch pipes to the backs of locals' souped-up pickups and dirtbikes and dragged them up the mountain. Using small power tools, they fit the pipes together and stuffed them in the dam. Finally, after twenty-four continuous hours of work, hacking through impenetrable thickets of bamboo, they had connected the full pipe system, changing the stream's direction.

The enforced idleness during the worst of the rain from June 27 through early June 29 had bred some discord among the divers in the camp. Divers like Ruengrit and Reymenants considered the British aloof, since they spent most of their time hunkered inside their headquarters office with the red sign on the door reading UK RESCUE DIVERS. The Brits surveyed some of the sites with Vern but didn't socialize, leaving the other divers with the impression that Stanton and Vollanthen didn't want to work with the Thai Navy SEALs—or with them, either.

To some degree the accusation was true. Cave diving is mostly a solitary experience, determined by the confines of a cave. There is seldom space for tandem dives, and even rescuers like Stanton and Vollanthen, who have known each other for decades, only swim side by side when a cave is wide enough to safely accommodate two divers abreast. The rest of the time they basically swim in single file, but minutes apart. And as the British Cave Rescue Council vice chair Bill Whitehouse said, "People have accidents and general miscalculations. If something goes wrong, it's a very unforgiving environment. There are many things you cannot put right, and that's the end of it. In a cave-diving accident you are either alive or you're dead." Stanton and Vollanthen are not dead, he explained, because "they are not adrenaline junkies, and try to stay calm and collected all the time."

The two Brits had only reluctantly hung around camp. When a few days earlier they had asked the British Embassy to arrange flights, their Thai counterparts gently informed them that "it wouldn't look good" if they went home so quickly. Weerasak, the Tourism Minister and the man who'd called Rob Harper (answering in his pajamas) at Vern's request and who had asked them to come to Thailand also personally requested they stay just a couple more days. So they did. When they heard about Reymenants's dive on Friday, June 29, the two Brits were impressed. He'd proven that guideline could be laid and that visibility had improved enough for some dive operations to resume.

On Sunday, July 1, with the rains tapering off,

Stanton and Vollanthen set off one at a time for Sam Yek—the T-junction. It wasn't easy going, but they got there and back without any significant problems, laying their own guideline to map the route from where Reymenants left off to about 250 yards past Chamber Three in the direction of the T-junction.

By Monday, July 2, early reports had it that the water in the cave had drained a foot. As with the earlier drops in water levels, no one knew for sure whether this was the result of that first effort to divert the stream, the pumps inside the cave, or simply the meteorological blessing of reduced rain. But now the pulsing currents from the Monk's Series had subsided even more than before, those yellow measuring sticks in the cave were getting visibly more exposed, and the tunnel had become significantly more navigable.

That afternoon, Stanton and Vollanthen suited up again. They had nearly a mile of diving ahead of them in a mission they reckoned would last eight hours. Everyone in camp knew that this would be a big day. At a predive meeting that Monday morning, the Brits, along with the Thai Navy SEALs and Ruengrit and Ben Reymenants's team, had decided that the foreigners would pave the way to the boys, but that it should be the Thai SEALs who made first contact. They spoke the boys' language, and it was important for Thai divers to be seen as planting the flag—the country needed the morale boost. Stanton and Vollanthen would lay guideline to an area about one to two hundred yards beyond Pattaya Beach. The rescuers knew the boys were not at Pattaya Beach, because the SEALs had been there during the first days of the search, finding only footprints. According to Vern's surveys and recollection, the next possible dry spot where the boys might have sought shelter was roughly three hundred yards from Pattaya, on a steep twelve- to twenty-foot slope high above the waterline. Ruengrit, acting as coordinator for the European divers, including Reymenants, told me, "We knew what the agreement was, and we all agreed to it."

That was not how it would turn out.

Chapter Nine
Contact

At the morning briefing, Stanton and Vollanthen
had been dispatched to lay guideline about 150
yards beyond the T-junction and keep pushing until
they ran out of line. They finished one 150-yard bag,
and then a second bag ended just before Pattaya Beach.
It wasn't much of a beach. The sand that the SEALs
had said they walked on in the early days of the rescue
had been submerged in water. They reckoned it was
Pattaya Beach because—according to the compasses on
their wrists—it was the only place that the cave jogged
eastward.

Kicking steadily onward from Pattaya Beach, they
began unspooling their reserve line—a much thinner
blue cord that ran about 250 yards. Again consulting
the compasses on their wrists, they were now headed

due south. They had committed to memory the little map that Vernon had sketched for them that morning, and knew this stretch was the longest, straightest passage in the cave.

When they finally sensed the air pocket above, they moved up toward it. It was easier to swim underwater, but they inflated their buoyancy vests whenever there was an air pocket—it was standard operating procedure for rescuers to investigate air pockets in case somebody was clinging to a dry ledge above. Stanton checked his air gauge—they were running low on the first third of their air—and they were also running out of guideline. Stanton estimated they had only a few yards left when he and Vollanthen started to kick upward to the surface. Low on air and line, it was likely time to turn back, thought Stanton.

But as soon as Stanton surfaced and took off his mask, he smelled them. He encouraged Vollanthen to do the same, and as the pair debated the source of the smell, they heard voices.

Coach Ek was too tired to move. He knew twelve-year-old Mick had the flashlight and asked him to investigate. But Mick froze; the dark creatures emerging from the water were the first things to disturb the absolute stasis of their existence in ten days. Not a breeze, not a bug, nothing at all, had intruded or changed in Chamber Nine—except for the progressive wasting away of their bodies. The refugee Adul, bold perhaps by necessity, grabbed the flashlight, headed to the water, and scanned it for those voices. He called to them in Thai, but quickly realized they were speaking English.

Following rescue protocol, Stanton and Vollanthen stopped on the far side of the canal to remove some of their diving gear—it would be safer over there. Sometimes "casualties," as they are called, get panicky and demand to be rescued *now*, clawing at the fragile dive gear. As they paddled across the flooded passage toward the boys and the overwhelming stench, Vollanthen pressed Record on the GoPro camera the Thai Navy SEALs had given him when he'd set off from Chamber Three, to document whatever it was the diver would find that day.

Following is a transcript of what would become one of the most widely viewed GoPro videos ever released. It begins while Vollanthen and Stanton are still in chest-deep water.

Vollanthen: Raise your hands.

Adul and a chorus of others: Thank you. Thank you. Thank you!!

Vollanthen: How many of you?

Adul: Thirteen.
Vollanthen: Thirteen?
Adul: Yeah, yeah.
Vollanthen: Brilliant.

In the middle of it you can hear the boys chattering among themselves in Thai.

Unidentified Boy: Do people know our bags are out there? Can you tell them about our bags out there?
Adul: Okay! I will tell them.
Unidentified Boy: I want rice.
Unidentified Boy: Can the rescuers come already?

The boys are sitting on their haunches, using their shirts to wipe the damp off their faces. They duck their heads to avoid the blinding beams of light. Adul keeps mustering a translation. He splutters out a sentence:

Adul: Backpack? Backpack with you go inside?
Vollathen: No, not today. Just two of us. We have to dive.

That last word, "dive," ends on an elongated upnote—explanatory but also deeply apologetic.

Adul [*in Thai to the boys*]: No rescue today . . .

Vollanthen: We are coming. It's okay. Many people are coming.

Adul: What day?

Stanton: Many, many people. We are the first. Many people come.

That promise would dog Stanton and Vollanthen with the sting of guilt over the next couple of days. They had no idea whether it was true. And the more they understood about the boys' predicament, the less optimistic they felt.

And then there was some confusion. Adul was trying to ask when their rescue would begin. Vollanthen, an ultramarathoner, seems to be breathing heavily. It's adrenaline rather than fatigue. He draws three quick preparatory breaths, like a father gearing up to assure his children it'll be okay when he knows it won't. The answer: "Tomorrow."

Stanton jumps in, assuming the skinny boy with the sharply protruding jawline is asking about the day of the week. "No, no, no, no," Stanton says. "What *day* is it?"

Unidentified boy: What day did they tell you that we can leave?

Fatigued themselves, and perhaps still jet-lagged, the divers are stumped. Vollanthen chuckles to himself, because he initially has no idea. Finally he says, "Mm—Monday. One week, Monday. You have been here"—he pauses to raise two hands and ten fingers, which obscure his camera—"for ten days. . . . You are very, very strong. Very strong," something the two men sincerely believed.

Adul and the boys had digital watches; they knew ten days had passed, but to hear someone else say it stunned him. He couldn't believe it was true. Shriveled from hunger and trying to process math and English at once, he struggled to mount a response.

Coach Ek snaps, "Who knows English, can you translate?" Tee says, listen to the foreigners. Adul responds that he cannot keep up with their fast-paced babble.

Vollanthen and Stanton are still huffing. Maybe it's from excitement. Yet what is most striking in that video is how composed the boys are. Rescuers are often concerned that their wards will try to grab onto them (which considering their plight seems entirely natural). But the boys hang back respectfully; they don't bum-rush the divers, so they decide to proceed up the slope. Vollanthen says to Stanton, "Okay, let's go up," and waving to the boys he says, "Okay, go back, go back, we are coming."

The boys remind their translators to tell the divers they are hungry. Adul says, "I've already told them!"

But then you hear the boys say in English, "We are hungry!" Vollanthen answers with genuine sympathy: "I know, I know."

The camera view skitters down as the men scramble up the mud-slicked bank. It is steeper and higher than they had imagined—a thirty-degree slope that goes up about forty-five feet. With the men struggling up to the boys' perch on the bank, you hear one of the boys repeat, "Eat, eat, eat, eat." The divers had not expected to find the boys and have no food to give them. And the news they have to offer isn't fantastic, either. As they make their way up the bank, they ask the kids to move back. The screen turns black, then we see a flash of the divers' shadows against the cave wall.

The boys ask when they'll be back. The divers hesitantly respond, "Tomorrow, we hope tomorrow. The Navy SEALs tomorrow, with food, the doctor, everything." They are fibbing again, knowing full well that only a handful of people in the world possess the experience to survive the round-trip journey to the boys, much less pull them out. Stanton thought to himself, *How on earth will we get them out? Diving anyone through that passage will be impossible.*

For the third time the boys ask, with a hint of desperation: "Tomorrow?"

Vollanthen assures them, "Yes."

Adul translates into the darkness: "An ambulance, a rescue truck will come tomorrow."

As Vollanthen's camera sweeps again to the right, you briefly see the team lined up on their haunches at the back of the cave. The two divers start to fiddle with their flashlights, trying to detach a couple from their helmets to give to the boys. Vollanthen mumbles apologetically to no one in particular: "My torch is quite shit, really."

Adul: I am so happy.
Vollanthen: We are happy, too.
Adul: Thank you so much.

That's when the video that was released cuts out.

But the divers would spend another twenty minutes with the boys. Stanton began inspecting their living quarters, easily finding the ten-foot-long "escape tunnel" they had been digging and the primitive sleeping area they had leveled out from the slope. He couldn't find the latrine, and given the smell, he hoped they were using a parallel passage. He also took stock of the boys, noting that the little ones and the coach seemed lethargic and frail. Some of the bigger boys appeared surprisingly energetic given their ten-day fast.

Hoping to ensure their message was properly conveyed, the boys began lifting up their soccer jerseys, revealing bony rib cages. Stanton had a Snickers bar in his pouch, which he never took out. It was just a single bar, and a dose of pure sugar could sicken them, he thought to himself. The divers and boys chatted some more. The plucky Adul asks the Brits where they are from.

"We . . . we are from England" says Vollanthen chuckling to himself that starving boys in a cave would still retain the curiosity to ask such questions.

As Stanton watched, proud of his friend, Vollanthen, the Cub Scout leader and dad to a kid just this age, began to lead the boys in cheers. He filmed them yelling "Goo, Thailand," "Goo, America," "Goo, Belgium," and so on. The boys grinned at the divers' enthusiasm, and he marveled at their spirit.

With additional promises to be back "tomorrow," as they gathered their gear to leave each of the boys came over and wrapped skinny arms around them. It was getting late, and the divers had been eager to set off, but for a moment they paused, letting the boys hold on to them for a beat or two longer, for as long as they needed. In a country where physical contact among strangers is unusual, and where a slight bow with hands pressed together in front of one's face takes the place of a handshake, the series of embraces showed the enormity of the boys' relief and gratitude. And the Brits were duly moved.

The British divers assumed Thai authorities were going to take over, and there was a chance—despite their promises—that they would never see the boys again.

"We were not in the position to get them out," Stanton would say later. "We didn't have a plan."

As it turned out, no one did.

It was one of the few nights that Major Charles Hodges, the USAF Special Tactics team commander, had returned to his hotel before midnight. He was settling in when his phone rang. His operations commander was on the line: "Hey, we found the kids."

Every few days a rumor would ricochet through camp that the boys had been found, so Hodges took a breath and asked his officer to go back and confirm it again. He didn't want to call his superiors—the colonels, who would call the generals, who would call the State and Defense Department bureaucrats—until they were absolutely sure.

"Because this is going to have global ramifications here," he tiredly explained to his sergeant.

Five minutes later the phone rang again. His tactical officer had walked over to the Brits' camp and met with the divers. "They absolutely found the kids. All thirteen are alive."

What surprised Hodges and many others was that the boys were all found in the same location. "I thought that they would probably have four or five in one spot, five or six in another spot. I certainly didn't think that all of them would find one location in the cave that was dry enough and large enough for all of them to fit and be there for that amount of time." When launching a rescue operation, it is far less logistically complicated to target one specific location than multiple locations.

The boys' group cohesion was no small piece of information. Hodges now knew that he was dealing with kids who had already aided in their own survival. To this point, they had done all the right things. They'd stayed together, found high ground, and had not tried to test uncertainty by trying to swim out of the cave. It told him these were kids who understood on an intuitive level not just how to survive, but how to survive together.

Cheers rang out at the ragtag camps of soldiers and volunteers that had sprung up around Mae Sai. Austrian Mario Wild from the Chiang Mai climbing team was staying at a nearby temple along with some American troops and others. After having spent the previous few days in the jungle, he had finally showered, the mud and sweat and bits of leaves pooling brown around his feet. Toweling off, he had just called his parents when a racket interrupted the call. Still wet, he wandered out to see what the screaming was about. People were elated. And after nine days of rain, mud, crappy food (at least up in the mountains), and suffering, people were also mentally and physically spent. Some couldn't get out of there fast enough. It was ten thirty at night that Monday as Wild watched one group after another clean off their gear, break down camp, and drive away.

"It was like, job well done. Let's go home," he recalls.

Minutes later Biw's English teacher, Carl Henderson—who couldn't have known that his little class cutup had tried to use some of the English he'd taught him—started seeing messages pop up on Facebook. When an announcement was made over the PA system in school the next morning, the kids jumped up and cheered and screamed.

Every day since they'd gone missing, the boys' twenty-eight hundred schoolmates from the Prasitsart school had sat cross-legged in that quad and prayed for the boys to be found safe. An instructor then led them through a brief meditation. On Tuesday morning, after watching that GoPro video of the boys' discovery, the instructor added an exhortation to practice physical fitness and English—using the Wild Boars as an example. One of the boys sitting there cried. It was Queue, the boy who had missed the cycling trip because he'd been up late watching the World Cup. Dribbling down his still babyish cheeks were tears of relief. A dozen of the jug-eared thirteen-year-old's teammates were in that cave, and he was supposed to be there with them. He had been tormented by guilt—he got to go home every night to his parents, his bed and comforting food, but they did not. At least now he knew they were alive.

The greatest jubilation, though, erupted in the little park ranger hut where the parents had been sleeping since June 23. Within seconds, cameramen had sniffed them out and poked their lenses through the open windows. Swept up in the moment themselves, the cameramen gave the thumbs-up. "Everything good!" they yelled to each other through the din. The celebration was even live-streamed. Parents stood up and cheered, high-fiving and hugging.

"They found them!" cried one of the parents, grabbing Coach Nok, whose head still craned forward in a neck brace.

The coach responded, "I'm very happy," barely looking up from his phone. He was searching for the stills and video that he'd heard had started to circulate online, asking the others, "Can I see the pictures? Are there any pictures? I'm so happy for you. This is amazing," he told the beaming mother of Titan, who, like her son, is smaller and seems younger than the rest.

A father, his face flushed from cheering and triumph, pumped two fists toward the camera; "I'm *super* happy," he told the cameras, showing equal parts victory and relief. "Thank you, all the news channels!"

Dom's mother, like them all, wept. Her family owns a corner jewelry shop in the local bazaar where thirteen-year-old Dom often helps out alongside his grandparents and sisters. The shop is famous for its amulets and most of the family members wear at least one or two of the plastic-encased figurines of the Buddha around their necks. Many of the amulets are blessed by Thailand's most revered monks, and the shop proprietors say they can bring health, virility, or general good luck.

The boys' plight had confounded her—what had they done to so anger the spirits? At some point during the sixth day after the boys had gone missing she had a premonition that Dom was no longer alive. She didn't dare tell the other parents or even her husband. But she simply *knew* the boys had gone too long without food. There wasn't a chance they were alive.

But now that was all behind her. That night, bathed in the cheers of fellow parents, she sobbed with gratitude to the spirits, to the divers, and to the Thai SEALs. And she allowed herself to savor the moment.

Tomorrow, she thought, *tomorrow, I will think about how they'll get them out.*

PART TWO

Chapter Ten
Suicide Mission

The discovery of the boys was now officially the story of the summer of 2018. The hundreds of accredited media crowding what had once been the cave's central picnic area now swelled to over a thousand. Reporters and correspondents were called in from all over the world. American networks like ABC News sent in massive reinforcements (including me). ABC News deployed resources from all over the globe, with producers coming from Los Angeles, Atlanta, London, Madrid, Hong Kong, an engineer from Los Angeles, and cameramen from London. We took up residence at a hotel about half an hour from the cave, eventually renting out a couple dozen rooms. Almost no expense was spared in covering the story.

At one point we had sixteen translators on the payroll. And as it was for the Special Forces, it became a journalists' convention.

Yet a glut of reporters does not necessarily equate to better information, which was hard to come by. The government had instructed the rescue teams, as well as the parents, not to speak to the press. So aside from a few confidential sources, most reporters had to rely on the press conferences, which often gave us little sense of what was really happening in the cave away from our muddy live positions.

As people around Mae Sai and the world experienced a mixture of relief that the boys were alive and anxiety about what would happen next, the diving camp quickly became a hive of activity—and tension. When Stanton and Vollanthen surfaced from the sump at Chamber Three, they'd immediately called out, "We found them!" The Thai SEALs were elated. They jumped up and excitedly grabbed the GoPro from Vollanthen, leaving the exhausted divers in the water; on every previous dive, those on solid ground had taken great care to help them up and out, but amidst their jubilation on this day they forgot the Brits. The Thai SEALs went to a corner of Chamber Three and conferred there. They immediately relayed messages through the walkie-talkie system to their commanders, and a few began scrambling out of the cave with the precious GoPro video.

In the time it took Stanton and Vollanthen to unhook their gear and set off on the swim, crawl, and hike back to the entrance of the cave, they were already being heralded as heroes—shaking hands the entire way and getting clapped on the back.

According to Stanton, by the time he and Vollanthen had reached the command center, still dripping water and sweat, the video had already been posted to the Thai Navy SEAL Facebook page and its two-million-plus followers—including nearly every member of the media. The funny thing was, the Brits had actually come to tell the Navy SEAL commanders to suppress the video. They feared it would put enormous pressure on everyone, including them, because they had no idea how they would get the boys out. But once they saw the global jubilation the video generated, they decided not to mention it.

The Thai commanders mined them for every speck of relevant information: what the tunnels were like, the boys' location, their health, their supplies, sleeping quarters, latrines—anything. They then thanked the Brits for their help and informed them that it would no longer be needed—the operation would now be in Thai hands.

As Stanton and Vollanthen soon learned, the Thai SEAL commanders, while happy the boys were alive, were apparently less than pleased with the Brits. In their postdive meeting on Sunday, July 1, and predive morning meeting on Monday, July 2, multiple sources tell me it was understood that Monday would be the big push—if they didn't actually find the boys, they would find evidence of where they had been or bread crumbs leading to where they had sought refuge. The conditions were right, the water was down well over a foot, and thick line had been painstakingly laid all the way past the T-junction. The best guess, using the existing surveys of the cave and Vern Unsworth's educated guesses about side rooms that offered high ground, was that *if* the boys had survived, they were likely a few hundred yards past Pattaya Beach, huddled in that small dead-end chamber with a high slope offering refuge from the water. Which is pretty much where they were found, in a spot that would eventually be known to everyone as Chamber Nine.

As the European divers, the Thai cave diver and GM manager, Ruengrit, and the other Thais understood it, the agreement among the divers had been that the SEALs would be the first to make contact with the boys—they had the language skills to communicate with them, plus it would reinforce the perception that this was a Thai-led operation. It presented an invaluable public relations opportunity that would boost national morale.

Ruengrit recalled that the agreement could not have been clearer: "We knew that was the agreement and we all agreed to it. After our team [Ben Reymenants and Maksim Polejaka] dove that day [early on July 2, laying line to an area just past the T-junction], the UK divers said they wanted to go in and pass Pattaya Beach, but instead of laying guideline, they used their own primary spool [what the Brits called their "reserve" line because they used it when they ran out of the larger rope] to go through from the last guideline point."

Basically, Ruengrit argued, because the Brits had used up all their thicker rope and deployed that reserve spool—which was far lighter, less cumbersome, and made for swifter progress—they had in turn made things harder for the divers who would eventually follow them. For less-experienced divers like the Thai SEALs, who would need to follow the Brits' guideline to the boys, the reserve line was much harder to work with. The Thai SEALs would need line robust enough to pull on—that thin blue line might as well have been dental floss.

For his part, Stanton called the notion that they violated any agreement "ridiculous." Everybody knew he and Vollanthen were the team who would likely find something, maybe even the boys, he said. They had never been specifically instructed simply to turn back at any specific point, and even if they had, they reached the boys before that blue spool of guideline ran out.

As Stanton said later, "Now that we found the route, what were they expecting us to do? Leave them the last ten meters?"

The concept of letting the Thais plant the flag with the boys made no sense to him. He and Vollanthen didn't work that way—they would keep pushing until they either laid all their line, hit a dead end, ran out of air, or found what they were looking for.

Regardless of which side remembered it correctly, the Thai Navy SEAL commanders were disappointed, and this discord over the Brits' discovery of the boys was just the beginning of the unease in the hours after the boys were found. Before the boys had been located, the different groups of divers—the Brits, Ruengrit's team, the Thai Navy SEALs, the Americans, the Australians, and the Chinese—had been largely aligned in their goal: lay as much line as possible to enable the farthest push into the cave. With the boys found, that goal had been accomplished, but now there was a bigger, more challenging question at hand: what to do next.

Ruengrit, Reymenants, and their team were back at their hotel in Pah Mee, giving an interview to my ABC News colleague James Longman, when the news of the boys' discovery started pinging on their phones. They stopped the interview—everything they had to say had just become moot anyway—got suited up, and went right back to the cave, with the ABC News crew in tow. They were surprised the Brits had found the boys and maybe a little envious.

"We laid the red carpet for them and they laid the British flag," Reymenants told Ruengrit.

By the time Reymenants and Ruengrit arrived at camp, a group of divers, U.S. Air Force Special Tactics team members, and Thai SEAL commanders had already gathered at the briefing area under the jumble of blue tents housing the Thai SEALs' operations center. They crowded around a pair of white folding tables lit by a batch of fluorescent lights. The Brits sat on one side, the Thai SEALs, including Rear Admiral Apakorn himself, sat on the other; USAF Special Ops captain Mitch Torrel sat at the far end. On the table was a 360-degree camera, recording the conversation. The Chinese team hovered in the background, trying to follow the conversation through its harried translator; beside them stood a hedge of other translators—mostly young Thais with notepads straining to catch up and relay the dialogue between the English speakers and the Thais.

Vollanthen and Stanton told the Thai SEALs, and later the Americans, that diving the boys out would be "absolutely impossible." Stanton remembers telling the group that the diving portion between Chamber Nine and Chamber Three was harrowing even for them: close to a mile long, in water so cold and so dark that it would spook even experienced divers. There were numerous line traps that would lead rescue divers smack into dead ends. At this point the rescuers had all been told none of the boys could swim.

As if all that weren't enough, those spectral figures squatting in the mud they had discovered a few hours earlier were clearly weak and desperately hungry. In a rescue dive, they could just as easily die of hypothermia as by drowning. It just wasn't feasible. Given their experience with the four flailing workmen they'd rescued a few days earlier, the Brits stated flatly that under the current conditions an extraction dive—at least, one in which they would take part—was an impossibility. According to Reymenants and several others present, the Belgian tried to approach the British divers—offering an olive branch and an alliance for the sake of the boys—but they rebuffed him.

For his part, Reymenants refused to accept the concept of abandoning living humans in that cave, risk be damned. He echoed a more activist group within the camp willing to risk lives in order to save the boys, and he called for immediate action—*now*.

Vollanthen cut him off, saying, "We can't just go in and bring them out. They will have to stay a bit, they seem okay."

"Are you a doctor?" Reymanents snapped back. "Do you have medical experience? Are you qualified to discuss their medical condition?"

Vollanthen demanded that Reymenants stop talking to him. To the Brits, who were now deeply committed to the boys who had hugged them hours earlier, his fervor wasn't admirable—it was amateurish.

The exchanges grew heated, and Ruengrit had to pull Reymenants from the tent and into the night to calm him down. The Thai SEAL commanders were there, including Rear Admiral Apakorn, and Ruengrit feared things would get ugly. Ruengrit's crew left the meeting. He and Reymenants felt snubbed—after all, Reymenants had more dives under his belt and had laid more line than anyone other than the Brits.

People in the tent that night tell me that everyone was trying to be measured, while the Belgian, according to Stanton, was "being emotive. He talked one hundred miles an hour and didn't get anywhere. He was pushing to get the children out. 'You have to do *something;* you can't leave them in there.' He was generally being annoying. And he didn't have any answers." In the necessarily hyperrational world of cave diving, "emotive" was synonymous with "hysterical," and thereafter the Brits treated Reymenants as the camp leper. He would not step inside the cave again.

It was an ugly episode that continued to fester and would later morph into a distracting sideshow that neither Reymenants nor the Brits wanted. Ultimately the British team would concede that Reymenants was instrumental in paving the way for them to reach the boys, providing valuable assistance in laying a few hundred yards of guideline. Nevertheless, he was not someone they wanted anywhere around them, and they would work toward barring him from the camp permanently.

Disillusioned and slighted, Reymenants and Ruengrit headed for the Chiang Rai airport the next day—Tuesday, July 3. Reymenants was bound for the Philippines for that delayed family vacation and Ruengrit was going back to work at GM. They were being replaced by another team of Thailand-based divers, henceforth known to all as the Euro-divers.

Later that night, Stanton and Vollanthen were exhausted and got back to their guesthouse late. Not too late, however, to throw back a few beers, allowing themselves just a sliver of satisfaction. (It was actually Stanton who drank most of the beer, since Vollanthen rarely drinks.) Indeed, the world was now talking about these British divers, who doggedly maintained their anonymity—refusing interviews, even turning their backs to photographers. But in the camp, there were 144 Thai SEALs and ex-SEAL volunteers who'd been slaving away over the past few days, risking their lives; as some of them saw it, the Brits had robbed the Thais of their moment. The Thai SEALs planned to snatch it back in what would be the most harrowing dive attempted in the weeks-long rescue effort.

Early on the morning of Tuesday, July 3, without informing any of the other international teams, the Thai SEALs mapped out a plan to dispatch four divers to the boys. They would plant the flag.

Four Thai SEAL divers set out from Chamber Three. Details of their dive to the boys remain sketchy, but the plan was for them to return in about eight hours. None of the four SEAL divers had been that far into the cave before—in fact, none had been much beyond Chamber Three—which meant they had to swim fifteen hundred yards of unfamiliar tunnel without a guide. They had no official cave-diving training beyond the instruction-by-experience they had picked up in the previous few days—most of which consisted of the quick hop through the sump separating Chamber Two from Chamber Three, about one hundred times shorter than the journey to the boys.

The journey was apparently so harrowing that at one point one of the divers lost his mask. When they finally arrived at Chamber Nine, six hours after starting out at Chamber Three, two of the divers had used up nearly all of their air—they would not have enough to return. Nevertheless, twenty-four hours after Stanton and Vollanthen promised to be back "tomorrow," divers again pierced the stretch of canal that locked the boys in.

Again the boys heard voices and saw lights rake the side of the chamber. They crept up to the edge of the water to see if maybe this batch of divers brought food. Before the divers said a word, the boys' flashlights made out the Thai Royal Navy's emblems on their dive gear and wet suits, and their hearts surged with pride. As the divers approached, they called to the boys, their voices thick with machismo:

"Here you are, Wild Boars, we've come to help you!"

If the trip was nearly fatal for at least some of the crew of four, they betrayed none of that to the boys.

The boys helped the tired Thai SEALs up the bank and showed them their humble home. The divers had brought packs of energy gels for the boys, and also real hope that their ordeal was near its end. As they chatted, the boys tried to describe their experience. The Thai SEALs didn't say how long they planned to stay or specify when a rescue would begin, and the boys didn't want to press them.

Three hours after that first Thai SEAL contingent set off, another team settled into the murk in Chamber Three. This second group of Thai SEALs following up the first mission was comprised of a medical contingent for the boys, ferrying space blankets, medical supplies, a medic, and made-for-TV former SEAL Lieutenant Commander Dr. Bhak Loharjun. Dr. Bhak, as he's known, heads the Thai Third Medical Battalion. What had become known as one of the world's toughest cave dives would be Dr. Bhak's very first.

"At first," Dr. Bhak later said, "we thought the passageways would not be tough. But when we got in the water, it was so murky that there was barely a foot of visibility even with our helmet lights." Even with the guideline to lead the team of three, Dr. Bhak, a Thai SEAL medic, and another Thai SEAL frequently swam into dead-end inlets and had to awkwardly maneuver backward. The passages were narrow, and though they pulled themselves along the guideline laid by previous divers, the current was stronger than Dr. Bhak had ever experienced in open water—it was like swimming against a rip current.

Like all the others, he had set out from Chamber Three. And like nearly all the others, he hit a snag. In his mission briefing earlier, the SEALs cautioned him about the countless obstacles that could emerge from the darkness to entangle dive gear, such as wiring from the early days of the search and the cave's unseen teeth snarling from the cave wall.

All of a sudden, his mouth was full of water instead of air. His mouthpiece had been ripped out. "And then the face of my son and wife came up—this was it, the moment that I thought of life and death." By windmilling his arms he managed to get a hand on his regulator, which was invisible in the murk. It took longer to get a grip on his racing pulse.

These were anxious hours back at Thai SEAL headquarters. Hours passed. Then an entire night. The Thai Navy SEAL captain in charge of dive operations suspended all dives during the time his team was inside.

When Dr. Bhak and his team finally dragged themselves up on the bank in Chamber Nine, they found the boys and the four Thai SEALs who'd stayed from the day's earlier mission waiting with smiles. Dr. Bhak was initially concerned the boys had been traumatized, that they might be physically and mentally incapacitated, but the Wild Boars surprised him. They were stronger than he'd expected and didn't seem to exhibit evidence of mental trauma. Dr. Bhak and the SEAL medic had orders to stay with the boys, along with two of the SEALs already there. The four had committed to staying with the boys, for the duration, come hell or high water.

Early on the following morning of Wednesday, July 4, the SEALs who were not staying made their way back to Chamber Three. When they arrived they were clearly shaken, describing to their commanders the horrific dive—those squeezes, the current, the eternity of darkness. They couldn't imagine bringing children through an ordeal that had expended every reserve of courage and strength of elite navy divers. The Thai SEALs now understood that a rescue dive was out of the question, at least for them. But the trio of divers also brought back tantalizing information from the boys and a new GoPro video. The video was rushed out of the cave and onto the Thai SEAL Facebook page.

The parts of this GoPro video that were released are different from the video shot by the harried Brits, who were more interested in capturing proof of life than a portrait of the boys' condition. The new video was shot in close-up—an intimate peek at the boys—and was meant to offer a booster shot to the morale of a worried nation.

The video instantly went viral—there were the smiling boys, and there was Dr. Bhak, robust and handsome, cheerily cleaning their foot wounds. The video begins with a pan, left to right, as Dr. Bhak dabs the boys' minor cuts and infections with ointment and iodine. The boys had suffered injuries similar to those of their army of rescuers—mostly small infections to the hands and feet. At one point the camera again pans from left to right as the boys say their names and clasp open hands together just under their nose, nodding their thanks to everyone. This video was a proof of life, but it was also a perhaps-inadvertent snapshot of their physical distress. The first boy you see is Mark. His face is skeletal, the weight loss magnifying his ears and a chin so sharp it seems poised to pierce his skin. And even before the camera pans away to other boys, Mark's smile quickly fades.

That pan reveals two sleeping boys, so weakened that they seem oblivious to the commotion and the cameraman's pep. One of them is little Titan, who blinks awake but makes no move to address the camera.

Nick gamely but wearily flashes two bony fingers in a peace sign. As the camera continues its pan you hear the rustle of kids moving in space blankets pulled up to their chins. Some wear them like skirts around their legs. Then there's Biw, all the chubbiness from those mid-lesson snacks in English teacher Carl Henderson's class melted away. Many of the boys are so cold they've pulled their filthy jerseys over their knees—exposing patches of red polyester burnished brown by mud. There's a moment of levity, though, when Mark thanks the press from around the world and everybody laughs.

The world was now watching even more intently—gripped by the incredible saga and the suspense of it all. How *would* they get the boys out of there? Millions of viewers toggled between channels broadcasting their favorite soccer teams competing at the World Cup in Russia and news outlets covering the plucky soccer team in Thailand. Players sent support, including Brazilian legend Ronaldo, who told CNN: "It's terrible news, and the world of football hopes that someone can find a way to take these kids out of there."

Those Thai SEALs who straggled back from Chamber Nine with the video of the boys may also have carried back hope. As soon as they arrived, still in their wet suits, they relayed to their commanders what the boys had told them: during their ordeal they had heard roosters crowing, dogs barking, and children playing. That morning, Wednesday, July 4, the commanders brought one of the exhausted divers to the American tent, and he again relayed this intelligence to a bigger group, including the USAF Special Tactics team and the British divers. Major Hodges and Master Sergeant Anderson were called in. Brits Stanton and Vollanthen were also there, busy stuffing waterproof bags with food they were going to swim in to the boys. Anderson questioned the Thai SEAL about what he'd heard; the Thai SEAL and his commander insisted that that was what the boys had told him. When Anderson and others asked the beleaguered commando whether he himself had heard anything remotely like humans, chickens, or dogs in the hours he had been in Chamber Nine, the Thai SEAL answered in the negative. The Thai SEALs' operations commander then asked the British divers who had just returned from Chamber Nine thirty-six hours earlier if such a thing were possible.

"One hundred percent impossible," said Stanton, explaining that the boys were six hundred yards below the surface and not in an area of any habitation. "It was almost laughable. Obviously we tried not to laugh," said Stanton later, adding that it wasn't funny that the boys said they had heard things and completely legitimate for the SEAL commander to ask about them, but he explained that the notion of a side tunnel that big was simply preposterous.

Hodges wondered aloud whether maybe there was something to it. Stanton looked at him cockeyed, knowing that the major had been awake for over twenty-four hours. Hodges and Anderson privately consulted Vern Unsworth and the British divers. Stanton and Unsworth explained that, were any such shaft to exist, it would have to be at least five hundred yards deep—a depth which could fit the Empire State Building with room to spare, and wide enough to base-jump into—so big that it would easily be spotted by a satellite. Anyway, everybody inside Chamber Nine would know if there was such a shaft, because a tunnel of that size, capable of carrying those sounds at such a distance, would deliver a gale of draft—since caves breathe, the bigger the shaft, the bigger the breath.

Satisfied by this explanation, Anderson and Hodges dropped it. The Thai SEALs didn't, though, immediately dispatching more teams to the mountain above, trying again to find that possible shaft. The Americans declined to send a team up the mountain yet again.

Word spread quickly through camp that the boys were able to hear sounds of chickens and children. The media ran with it, raising expectations of an as-yet-unrevealed alternative entrance, one the spirits of the cave seemed to have hidden from the many hundreds of troops stomping all through the Sleeping Princess's forests.

Despite the newfound comfort of knowing Thailand's elite were with them till the end, the Thai SEALs' arrival in Chamber Nine presented new challenges. In addition to the twelve boys and their adult coach, there were now four more grown men in the chamber with the boys—the equivalent of adding ten more Titans or Marks, who each weighed about seventy pounds. They were rapidly consuming the dwindling amount of remaining oxygen; like the boys, the SEALs would need to be provisioned with food, blankets, flashlights, and—ultimately—dive tanks, along with at least one new mask. The international teams struggled to understand why the "comfort dive," as some called it, was not coordinated with the other dive teams, and why the Thai SEALs brought so little food with them.

Unwittingly, and to the head-wagging dismay of the international teams, the Thai Navy SEALs had significantly complicated the mission to resupply and rescue the boys. Instead of adding time to the mission's clock, they had sped it up.

Chapter Eleven
The Zero-Risk Option

And then, after the Brits went in on July 2 and the Thai SEALs on the following day, dive operations skidded to a halt.

The Thai SEAL ops commander took to heart his commandos' warnings about the hazards of the dive to Chamber Nine. He suspended any further Thai SEAL diving expeditions to the boys—they had made their point and risking additional lives was unnecessary. Anyway, the Thai *pu'yais,* as the political bigwigs are known, preferred waiting it out—drilling a relief shaft, finding another entrance, waiting for the monsoons to clear, anything that would not risk the lives of the twelve boys. And the Thai SEALs were not alone: pretty much everyone, including the Brits, considered a dive mission under the present conditions too risky.

The Thai SEALs started thinking about the long-haul survival of the soccer team, digging right back into their bottomless supply chain. By now, in addition to the sleek Embraer jets, lumbering C-130 cargo planes were delivering mountains of donated supplies, including over four hundred gleaming aluminum-alloy air tanks and half a dozen refrigerator-size compressors to fill them with. The rescue team had oxygen compressors in case they wanted to adjust the air mixture by jacking up the level of oxygen. They had dozens of inflatable buoyancy control vests, lead weights, belts, webbing. There was so much gear they had to erect a special tent to house it all. In addition to stuffing the cave with air tanks and setting up a high line in both Chambers Two and Three to more efficiently zip the tanks down to the divers below, they ordered miles of thin hose. If they couldn't bring the boys out into the fresh air, they would pump fresh air in to them.

They already had the oxygen compressors. Now trucks began unloading thick spools of air hose right in front of the curious press, which was quickly informed of the audacious plan. Workers in orange overalls became human spools, clasping their hands together with the coils of hose unrolling on their forearms as Thai SEALs dragged them into the cave. But the farther the hoses were wormed inside, the slower the progress.

Complicating the mission was the attempt to attach a telephone line to the oxygen hose. The thinking was, if we can get air to them, we might as well get them a link to the outside world. At this point, the only way of transferring information from Chamber Nine to the outside world was by diving it out. The phone line would enable the boys and their parents to speak for the first time in nearly two weeks. It would enable the Thai SEALs to receive orders from their commanders and to more precisely coordinate the supply chain. It would lift morale. It could make a four-month stay in a foul tomb more sustainable.

The UK dive team considered the notion of running an oxygen tube through a tortuously contorted, flooded tunnel lined with shark's-tooth rocks a fool's errand. It might have been possible if they had unlimited time and resources and the cave was dry, but there were now seventeen people in Chamber Nine. Past Chamber Three, physically pushing the manhole-wide coils through obstructions barely half that size would be impossible, and possibly fatal. (If you have ever tried to pull a garden hose through shrubbery or a few rocks, you might understand how difficult an undertaking this would be.) Dozens of segments of hose would need to be screwed together. Even if they managed to thread it to the boys, maintaining each valve and connection would require an army of scuba-diving technicians.

The compounding number of tanks necessary to complete and maintain the hose would have sapped the supply chain. It would also have compounded the level of risk—packing more divers into places they were untrained and unequipped to go. The Brits weren't chafing at this plan because it was unfeasible; they were uninterested because it required risks they felt were unnecessary. It's one thing to take a risk when you believe in the end result; it's another when the end result looks like failure. The Brits were the world's top cave-rescue divers partly because they had an actuary's knack for calculating risk, so when asked to help they politely declined.

While the Thai SEALs forged on with the oxygen hose plan, the Brits maintained, along with the U.S. Special Tactics team, that the first order of business must be to supply the boys with the food they would need to survive the next few days, and possibly the months of the monsoon season.

So on Tuesday, July 3, as the four Thai SEALs were winding their way to the boys, Anderson had canvassed Stanton and Vollanthen—who had seen the boys' miserable living quarters—on what the boys and their coach might need. Items on the list were things like food, blankets, headlamps, and batteries. They started rummaging through the camp to scrounge up what they could. They found five headlamps and fifty extra sets of batteries. They found water pumps so the boys and the Thai SEALs could drink clean water. That would buy time. Getting them food would be trickier.

Most food, especially anything that is packaged, is buoyant—it floats.* Since hauling anything with substantial volume would severely limit the calories ferried in, Anderson had an idea. His team had brought more than one hundred military rations called MREs—meals ready to eat. Each package is designed to sustain a 180- to 200-pound soldier in strenuous combat situations and can go unrefrigerated for five years. The meals average about 1,250 calories; among the more highly caloric items is the pork patty meal, with 1,345 calories, 82 grams of fat, and 46 grams of sugar. Being packed with calories, MREs were just what the boys needed. And since the food provided at camp by the volunteers had been both delicious and plentiful, the Special Ops team had until now no reason to dig into their MREs.

---

* There were reports that divers managed to get them sticky rice, pork, and milk—just one of dozens of false reports about the boys' time in the cave.

Anderson enlisted some of his men, Stanton and Vollanthen, and even the pair of Tourist Police officers who had been assigned to serve as the Brits' minders to strip down the MREs. They chucked the near-atomic-bomb–proof plastic wrapping, the Tabasco sauce, the gum, the matches, the napkins, the "flameless ration heater," and anything else that was either not caloric or that might float. They were left with the vacuum-packed main courses—flavors included beef ravioli, beef taco, apple maple oatmeal, pot roast, garlic chicken, and chicken fajita—and desserts. Said Anderson, "Yeah, we figured they could use those too, why not?"

Anderson had calculated that if each person in Chamber Nine ate just one of those full meals a day, it would sustain the group for about a week. At least it would buy them time. They then tried to stuff 117 of them—every single one they had with them—into homemade, neutrally buoyant tubes the UK team brought with them. The tubes looked like missile casings—three feet long, a foot wide, with steel O-rings on either end and weighted down with lead. They had so much food and gear that they needed additional bags. So they borrowed three of the Special Ops team's dry bags (giant waterproof duffels), crammed them with as much as they could, squished them to get the air out, and hoped they wouldn't float too much. Stanton's tube was neutrally buoyant, meaning it would theoretically stay wherever in the water column the divers took it—but that giant dry bag wasn't.

"No problem," said Stanton. When he got into the cave he asked his Tourist Police minders to grab a few rocks and a few handfuls of sand. "That'll do," he announced, dumping the dirt and rocks inside as the Americans stared wide-eyed. Stanton shrugged. "That's what we do." Cave divers sometimes rely more on resourcefulness than planning. Vollanthen carried two of the sand-laden bags.

Shortly before the Brits entered the water that day, rescuers and pump workers began scurrying around outside the cave. A Thai SEAL was brought out with some help after nearly drowning. It apparently happened in the sump between Chambers Two and Three, which had shrunk, leaving a submerged section a little longer than an average swimming pool. The foreign divers were confused—how could a scuba diver nearly drown in such a short section?

The answer was: when he's not using scuba equipment. For the first time since June 26, the sump between Chambers Two and Three had shrunk to the point that a free diver could take a few deep breaths, plunge in, and duck-dive his way across—pulling himself on the guideline when necessary. That this was possible didn't mean it was advisable. The U.S. team and the Australians—who had discovered they were simply too big to fit through the manhole-size hole after Chamber Three and had started helping the Americans with logistics—continued to use small tanks for this short dive. But one of the Thai SEALs, who was freediving the section, apparently became ensnared in some of the hoses or electrical wires running between the two chambers. A limp body bobbed up to the surface. Divers and U.S. pararescuers who happened to be there jumped in and pulled him out—he was unresponsive and seemed dead, but they pounded on his chest, tilted him to clear his airway and purge water, and after a few minutes of intense CPR they managed to revive him. A day later he was back on the rescue.

Nevertheless, the Thai SEALs forged on with the oxygen pipe plan. They didn't really have a choice: military leaders had been talking about it as a fait accompli. A Thai junta spokesman, Major General Chalongchai Chaiyakum, told the press, "When the telephone line is ready, we will have relatives talk to them. The pressure will be immensely reduced." This statement was completely divorced from the reality inside the cave, and served only to increase the pressure.

Thai political leaders had promised the world a

"zero-risk" option to rescue the boys, which meant either waiting out the monsoons, drilling a relief shaft, or finding the mythical alternative entrance. The only way to ensure they didn't die of asphyxiation before November or December, when the rains stopped or the drills found them, was snaking an oxygen tube to Chamber Nine.

The "zero-risk option," however, referred only to the boys. In a photo of one of the planning sessions, just to the right of Rear Admiral Apakorn, is First Petty Officer Saman Gunan, examining the plan on the whiteboard. He was to be one of the divers helping to install that oxygen tube. Within thirty-six hours he would be dead.

On July 4, shortly after the Thai SEAL almost drowned in the sump between Chambers Two and Three, Vollanthen and Stanton dipped back into the water at Chamber Three on a mission to mule calories to the boys. As always, Anderson and Hodges knew they wouldn't hear back from them for six or seven hours. Despite the lower water level, the current remained powerful enough that a leisurely swim would mean not moving. Between the current and the fact that they each schlepped dozens of pounds of gear that kept hitting snags, the going was slow. Their legs burned from kicking, and their hands were cracked and sore from pulling on the guideline. Their heads kept knocking dangling stalactites that they simply couldn't see in the Coca-Cola-colored water.

Vollanthen is an ultramarathoner and Stanton has kayaked nearly nine hundred continuous miles around the island of Tasmania; still, by the time they reached the boys, they were heaving in their regulators. Vollanthen's duffels were extremely cumbersome. He made frequent stops for ten or fifteen minutes—pit stops during which he told Stanton he couldn't go on and had to drop one of the bags. It was Vollanthen's most grueling dive, but he pushed through with the duffels.

When they finally arrived in Chamber Nine, "it was quite a ceremony," as Stanton recalled. Foreign as it was, the mushy stuff in the aluminum packets was the first real food the boys had seen in twelve days.

This time the divers' stay was shorter. Vollanthen and Stanton delivered another series of messages to the SEALs. They only stayed for twenty minutes and didn't see the boys tuck into the food.

In addition to the meals, Stanton and Vollanthen carried in two bubble-wrapped palm-size packages and a letter to the boys and the SEALs. The SEALs received instructions, and the boys were asked again, this time in writing, to describe precisely the sounds they had heard. The SEALs were not going to give up on the notion of finding an alternative route out, as unlikely as it seemed. To that end, one of the small packages Stanton and Vollanthen brought in was an HTC phone, which the SEALs back at camp had asked the Brits to deliver to the SEALs in Chamber Nine in the hope that they might be able to connect to a satellite positioning system. (According to Stanton, such devices exist, but none are powerful enough to penetrate six hundred yards of rock.)

The other bubble-wrapped device came from the Americans. It was a confined-space air-quality monitor. Anderson had asked Stanton to take a measurement, but also to verify the readings: "Can you take video of it, too? I need to have exact proof of this, especially if we are going to show this to the decision makers," Anderson told him.

Stanton did as he was told. The reading was 15 percent oxygen, and the meter began flashing red and beeping. The U.S. Occupational Safety and Health Administration defines anything below 19.5 percent to be hazardous. Oxygen levels between 15 percent and 19.5 percent cause "decreased ability to work strenuously" and "impaired coordination." At levels below 14 percent, respiration increases, and people start to get loopy and their lips turn blue. While the 15 percent reading in the cave set off alarm bells, Stanton maintains the reading might have been erroneous. He said the meter is designed to be calibrated first outside a confined space. But because they dove it in, Stanton had to calibrate it inside Chamber Nine. Given his experience mountaineering and in other low-oxygen environments as a firefighter, he surmised that the oxygen level was likely a little higher than 15 percent. Whether the reading was precisely accurate or not, it highlighted an emerging and immediate problem: there was an ever-dwindling amount of air in the chamber. Suddenly everyone began to focus anew on this problem of air.

Whatever the exact oxygen level in Chamber Nine, it was lower than ideal for seventeen people. Still, oxygen deprivation presented a possible explanation for the boys' insistence that they heard roosters and dogs. After eleven days of fasting, the boys' glycogen levels would have dropped, and low blood sugar could have led to low oxygen in their brains, which is associated with hallucinations. Or maybe the hallucinations were caused by continuous low levels of oxygen in Chamber Nine, exacerbated by dipping blood sugar.

For his part, Stanton says neither food nor oxygen had anything to do with it.

"It's absolutely normal to hear voices and sounds in

caves. It's not even unusual for experienced cave divers, because the cave plays tricks on you," he said. During his own expeditions in caves, Stanton had found himself turning to greet straggling cavers chattering behind him when his headlamp only illuminated an empty passage. Counterintuitively, it's the silence that gins up the sounds. A cave is undisturbed by any of the ambient sounds that populate our lives: the rustle of leaves, lawn mowers, distant buses, dogs barking, the beeps of delivery trucks backing up, toddlers' whining for attention. Therefore, what our brains would normally filter out as background noise in our world outside is amplified inside a cave. Each plip-plop of water colonizes one's auditory senses—sometimes, maddeningly, it's the only thing cavers can focus on. Those sounds have triggered panic attacks or even temporary madness among cavers. The boys might indeed have thought they heard something, and they may all have heard similar things.

"That doesn't make them crazy," said Stanton. "It just means they've spent too much time in a cave."

In his daily press conference that afternoon Governor Narongsak reassuringly informed the world that the boys were safe inside Chamber Nine and that "there was no need to rush anything." Narongsak himself told me later that he'd been averaging less than two hours of sleep a night, and he truthfully told reporters that day that "no one has rested since day one." But then he added: "We hope that the telephone line will be completely installed by tonight. There's nothing to be concerned about for the moment." Except there really was. Because that telephone line would never be completed, much less in a few hours, and the risks to the boys were piling up.

U.S. Air Force meteorologists were feeding Anderson and his team the latest weather data. It didn't look good. The satellite map showed what looked like a string of cotton balls headed their way: monsoons. The forecasts had been unreliable thus far, but monsoons can easily dump nearly a foot of rain a day, and everyone knew significant rainfall of even a few inches would force them to suspend operations. The pumps were barely holding out as it was. So Anderson started doing calculations: a meal a day for seventeen people for more than three months. Somehow he'd need to build a cache of eighteen hundred meals over the next ten days if the boys were not going to starve to death during the monsoons.

The Americans knew they would need more divers if any attempt at resupply was going to work. Feeling this urgency, they asked the Thai SEALs to start calling in the cavalry—any experienced cave diver. "Okay, call them in" was the response. And they scoured the camp for skilled divers among the many hundreds of personnel there.

After the commotion of the near drowning on July 4, the Americans walked over to the Euro-divers, the group of European expats living in Thailand who'd come to replace Ruengrit and Reymenants's team. "Euro-divers" was a bit of a misnomer, but it stuck. This group consisted of Danes Ivan Karadzic and Claus Rasmussen, German Nick Vollmar, Finn Mikko Paasi, and Canadian Erik Brown. They were a motley bunch. Karadzic, Rasmussen, and Paasi are in their mid forties. Paasi and Brown had shoulder-length dreadlocks. Rasmussen, Brown, and Karadzic had arrived days earlier, but were told by the SEALs to sit and wait for an assignment. After the Reymenants episode, the SEALs, not wanting to agitate the Brits and unsure how to utilize the Euro-divers, ended up essentially benching them. For two days, until the American team's Captain Mitch Torrel started chatting them up, the Euro-divers sat there in their wet suits, ready to dive, while the action unfolded around them.

Captain Torrel asked for their CVs, saying simply, "We need divers." The American then began peppering them with questions: What kind of gas would they need? What kind of gear did they use? And as Karadzic remembers it, "they asked us again and again, almost as if they couldn't believe us, if we were actually willing to go in." They were.

Rasmussen, an instructor well known inside the insular cave-diving world, was their leader. He said the Aussies and the Americans "were trying to figure out what the hell was going on and what do we do now." Rasmussen and Torrel, a pillar of a man who'd been a standout on the Air Force hockey team, got to talking. Rasmussen told him, "The Thai SEALs are following stupid procedures. Maybe they should use people like us if they are trying to do this right." To Torrel that made sense. By the end of that day, Major Hodges and his team had officially recruited them; henceforth they would be attached to the U.S. Air Force team.

On that evening of Wednesday, July 4, as the reporters trailed off, the workers at the food trucks started to bag and refrigerate the day's mounds of rice, and mosquitoes began to feast on the bounty of humans still there, Rasmussen sat in on his first meeting at headquarters with the Americans, the Aussies, and the Brits. Stanton and Vollanthen had just returned from delivering the food and taking those oxygen-level readings in the cave.

The mood was grim, recalled Rasmussen. "Most of us sitting there were talking on a non-bullshit level, and I was very much agreeing with the Brits. Since the Aussies and Americans only had one guy who was cave trained, we needed more divers. So we started coordinating what we could feasibly do."

Because of the weather forecast, the MRE resupply became a focus. Anderson figured he could get his hands on eighteen hundred MREs pretty easily. After all, the U.S. Air Force was sparing no expense on this mission—when a few days earlier Hodges and Anderson asked for that confined-space air-quality meter, the Air Force flew the palm-size gizmo and some extra gear in a C-130—a plane big enough to carry a tank. The question was less whether he could get the MREs to the Thai cave in time, and more whether he could get the MREs to the boys in the cave.

Stanton and Vollanthen had ferried in more than a hundred meals on their mission that day, and to those who saw them that night it was clear the trip had utterly exhausted some of the most fit and skilled divers in the world. Each meal weighed about half a pound. There were only a few divers in the world capable of completing the difficult journey, and supplying the boys would require about twenty trips—a hell of a lot of dives. The delivery divers would have to be completely self-sufficient—there were very few emergency tanks stockpiled beyond the sump after Chamber Three.

There was also the risk that inexperienced cave divers brought in as "meal mules" could end up stranded with the boys and in need of rescue themselves—making more mouths to feed and more lungs to consume the limited oxygen.

The math just didn't work. There would be no way of fully supplying the boys before the rains. Even as the Thais plodded on with the oxygen line, the foreign teams understood that it would never reach the boys in time, nor would the massive resupply of MREs. Hodges and Anderson concluded that they had no choice but to scrap the resupply. They had much less time than they thought. There was also some concern among the rescuers that if those eighteen hundred meals arrived, the Thai authorities would press them into delivering them to the boys, risks be damned. Quietly, the teams decided not to formally request those extra meals. All this, plus their fatigue, contributed to the Brits' overwhelming pessimism about a resupply mission.

Late that night the Americans, along with Stanton and Vollanthen and the Euro-divers, started throwing around ideas for a possible rescue dive. Assuming for the moment that a dive rescue was a necessity, they thought they could scrounge up thirteen divers who had the skill to actually get to the boys and bring them back. They debated doing it in a single day, the so-called grand-slam option: a long convoy of divers and their charges, one diver per boy. Given the boys' weakness, the unpredictable cave conditions, and the new divers' unfamiliarity with its jagged, seemingly booby-trapped labyrinths, the team calculated an 80-percent fatality rate—which would likely mean ten dead children. That attrition rate presented additional pitfalls: most of the Euro-divers were parents, some of whom, like Rasmussen, were bringing up children in Thailand—would they, as Rasmussen put it, "be willing to bring out dead kids, given that we live in Thailand?" For the right plan, Rasmussen answered, "I told them I was willing to do it." It was a remarkably courageous decision for the fathers in the group, who possibly faced not only a lifetime of guilt but also the wrath of an unpredictable military junta.

Vollanthen is also a father, and he quietly told Anderson that he didn't know if he could deal with the guilt of having a hand in the deaths of children—even if it came in the process of trying to save them. The Americans told him, "John, you don't have to do this, you're just a volunteer. We understand that. But wouldn't you want to know that you tried and gave it your best shot? Because if you and Rick don't lead this mission, you can pretty much guarantee that they are all dead if you leave." It was enormous pressure, but it was likely all true as well. Vollanthen and Stanton stayed, and committed themselves to saving however many kids they could. But it would have to be done under very specific conditions.

Every plan they discussed that night seemed to offer an unacceptable probability of fatalities, including the grand-slam option. There was one controversial building block upon which the Brits insisted that every new idea had to be built—the boys had to be fully sedated. After their brief experience with the four flailing pump workers they'd rescued a week prior, Stanton and Vollanthen insisted they would not participate unless the "casualties" were completely inert. Binding the kids and swimming them through without sedation would have been terrifying for the boys and possibly dangerous for the divers, so the only option to mitigate possible trauma and enable the divers to do their job was knocking them out. Among the little group there was nearly a century of cave-diving or caving experience, but none of them could think of any precedent for such a rescue.

Regardless, they needed more divers. Two of the Euro-divers, Nick Vollmar and Mikko Paasi, were on their way in from Europe. The UK dive team also called in its own reinforcements, including Jason Mallinson—the Jason Statham look-alike who had been involved in that Mexico cave rescue with Stanton in 2004—and another rescue diver, Chris Jewell. Jewell was an expert cave diver and technician, but this would be his first rescue operation. They also called in three other support divers through the BCRC: Connor Roe, Jim Warny, and Josh Bratchley. Stanton felt comfortable with the team he was assembling; he had worked closely with all of them aside from the Eurodivers, whom he was growing to trust.

He also knew who he did not want on his team: the Thai Navy SEAL divers. He figured they would have been unable to communicate with the rest of the team, plus their recent mishaps proved them to be courageous but technically lacking—they just weren't trained for cave diving. His primary concern, however, was his own (and his team's) survival. He began with the assumption that if the Thais joined the effort they would send a massive team—more than was necessary. They would likely not be posted to the far reaches of the cave, which meant they would be somewhere near Chamber Four—the only way out. If the cave started to flood, he assumed that the Thais would react as any novice cave diver would—scramble for the exit. And since getting to the exit meant shinnying single file out of that manhole-wide 150-yard sump between Chambers Four and Three, a human logjam would stack up.

The last thing he wanted was to be what he called a "tail-end Charlie" at the back of that line of divers.

Because sedation was nonnegotiable, the linchpins of the team would be two Australian divers, neither of whom was currently on-site: veterinarian Dr. Craig Challen and anesthesiologist Dr. Richard "Harry" Harris. If the elite British rescue team was a rare breed, then Dr. Harris in particular was a unicorn: one of the world's most experienced cave divers, who happened to have participated in previous rescues, and who happened to be an anesthesiologist.

They were quietly put on standby, at the request of the Brits. Later that night Stanton started to consult Dr. Harris via text. He recalls, "I was talking to Harry [Harris] about how to go forward. [The team] had agreed that the sedation plan was the only conceivable option," and that in terms of diving they shouldn't count on the Thais in the far reaches of the cave.

Stanton wrote to Harris, "The Thais aren't going to do anything, would you consider sedation?"— meaning, would Harris consider administering the sedation?

Harris replied, "Sedation not an option." Meaning that he was not willing to be the doctor who anesthetized twelve boys and their coach, the majority of whom were expected to die on the way out. If things went south, it could land him in a Thai prison or, at the very least, pave the path to quick career suicide.

Stanton left him with a last message—something to think about overnight: "If not sedated they are not coming out."

That night, Wednesday, July 4, Derek Anderson finally slept in a bed. He came back the next morning for a meeting with the divers and his team.

"As with all big problem sets," he told the team, "you have to stop and take emotion out of it." The divers had also gotten some rest and Anderson in particular was ready to attack this unique problem set anew.

"Okay," he told them that morning, "you don't think it's feasible. But if we were to attempt it, what would this look like to you?"

They mapped out the massive manpower available, the number of air tanks already staged in Chamber Three, and the number of experienced divers needed. They took inventory of the gear they had, including the U.S. team's four positive-pressure scuba masks—which pump a continuous flow of air into the mask, unlike other full-face masks, which provide air on demand and would prove essential for the rescue later on. They started scratching out new ideas. The removal en masse of all thirteen in a day was the first notion they rejected.

Anderson figured with the right plan, proper staging of equipment, and the right divers, it was possible. But only if *both* the weather held and the Thais budged from their demand of a "zero-risk option." Any rescue plan meant the possibility of casualties.

None of it was possible without an anesthesiologist who was a world-class cave diver. And since there was only one such person, Stanton again reached out to Dr. Harris, this time in "a proper phone conversation." They'd known each other for eleven years and shared the kind of trust engendered by watching each other's backs in the world's most unforgiving places—in the bellies of caves seven hundred feet under water. Stanton didn't have to sell him too hard. Dr. Harris had slept on it and told Stanton that it might be done, but friends had warned him overnight not to set foot in the country until he was guaranteed immunity from prosecution. Stanton started scrambling around the camp to find an Australian embassy representative to arrange the immunity through official channels.

That night I ducked under the caution tape dividing the parents from the rest of the camp and straightened up inside the parents' tent. The parents had been strictly forbidden to speak to the press. The government didn't want any more tearful interviews with parents tightly clutching framed photos of children they might never see again. The parents' now ever-present government minders looked up from their phones briefly. My translator and I told them we simply wanted to deliver a message. They saw that I was accompanied only by a translator and no cameras were in evidence, so they resumed their scrolling. I went over to Biw's mother, sitting patiently in a plastic chair under that blue tarp amid a cast of exhausted parents—many of them in surgical masks. A large flatscreen had been set on a folding table, and parents absentmindedly watched. But mostly they just sat there hollow-eyed. It was dark now. Biw's mother had a broad face, with high cheekbones and deep laugh lines from better days. She agreed to talk with us and ushered us quietly to the area behind the tent. The smell of runoff from the overflowing bathrooms nearby was ferocious. She was wary, but warmed up as soon as I presented my phone to her, with some pictures I'd taken of her son's artwork.

A couple of days earlier I had visited Biw's private school, a different one than Prasitsart. At this school the kids were also dressed in khakis, but with neckerchiefs tied around their necks, looking more like Scouts than British colonial officers. I chatted with the students and English teacher Carl Henderson, a former IT consultant, about Biw. The girls coquettishly giggled, covering their mouths at Henderson's gentle description of him as a popular kid but a bit of a ham—quick with a joke and just as quick to smuggle in bags of fried snacks. Biw also liked to doodle as Mr. Henderson rattled on about conjugations and tenses.

When I asked to see a sample of his doodles and artwork, a couple of the kids ran into the classroom and came back with Biw's notebook and a T-shirt, as if they were sacred relics. The children flipped through the pages—some attempts at English, a few math problems, and a bunch of perfectly symmetrical geometric drawings in Magic Marker. Biw had also drawn casts of cartoonish characters of his own design. The T-shirt that he'd made was populated by those same characters. The design looked like a city skyline, but dwarfing the spired skyscrapers were giant anthropomorphic fruits vaguely resembling the old Fruit of the Loom characters. There was a giant pencil and palm tree, a building-high skateboard, and the Eye of Providence—the wide-open all-seeing eye at the top of a pyramid you see on a dollar bill.

It shattered the vague two-dimensional image I

had of these lost boys. They were just kids, who day-dreamed and doodled, who had friends, who suffered through the tedium of classes so they could get outside for playground soccer. They were humans, who came back sweaty and stinky from running around, who acted tough in the schoolyard and then fled home to the bosoms of their mothers.

And now Biw's mother's face was lit by the glow of my iPhone screen. She'd seen all his drawings before, but not on someone else's camera. She looked up at me and the translator, smiling—her suspicion melted away, but not her worry. Maybe I had made it harder on her. I'm not sure. Moments later, one of the minders—realizing his ward had gone astray—called her in and forced us out.

The families had been kept on a need-to-know basis. Biw's mother knew, of course, that her son had been found, that he was safe for now, and that Thai SEALs were now hunkered down with the boys, come what may. But they were never informed about rescue plans ahead of time, much less consulted about their opinion of, for instance, waiting out the monsoons versus an immediate rescue. The SEALs were going to bring the boys out. The families had faith in them. And so they sat, waited, and prayed.

Chapter Twelve
Letters Home

Many of the workers and divers looked as though they had been punching rocks. Their hands were bloody and infected. On a given day a couple hundred people would go in and out of Chamber Three. A main traffic node, it had become the depository for hundreds of air tanks. It had also become a depository for urine and perhaps other bodily excretions. In addition to what looked like a messy fuse box of giant pipes, there was now trash building up and the smell of a never-cleaned highway rest stop. Normally cavers pee in bottles and take them out when they leave. But not here. There was too much work to be done.

Perhaps, then, it's unsurprising that it was unlike any cave Chris Jewell and Jason Mallinson had ever been in. Caves are typically solitary places. They are quiet and peaceful and largely untouched by humans. With the pumps whirring, dozens of voices in half a dozen languages pinging off the walls, the pounding of hand-held jackhammers, and the clanging of scuba tanks, this was far from peaceful. Back in London, Jewell had been acting as the British Cave Rescue Council's liaison with Stanton and Vollanthen in Thailand. After the two Brits located the boys on Monday, July 2, Jewell had run a marathon of interviews on Tuesday. At six the next morning, he and Mallinson received the call to come out. It would be Jewell's first rescue operation.

Both were a good fit for the job. Mallinson is a rope access technician, using his caving and climbing skills to access hard-to-reach places—like skylights in malls or the glass skins of high-rise buildings. He was just a pup when he started, back when he had hair atop his big impressive dome. He started caving at seventeen and then taught himself the basics of cave diving. He had been caving and diving with Stanton for nearly twenty years, and along with Stanton and Vollanthen had set a world record for longest cave dive back in 2010. Like many cavers, he is serious and intense. He almost invariably wears cargo pants and a T-shirt that had seen some tugging at the collar. Jewell is an IT consultant and looks like one: slightly nerdy, bespectacled, with a frank, open face. At thirty-five, he is younger than

Mallinson and Vollanthen by at least a decade, and twenty-two years younger than Stanton. But he was a rising star in the caving world, having led an expedition to the Huautla cave in Mexico, considered one of the deepest in the world—plumbing a spot called "one of the most remote yet reached on earth." The fifty-seven-year-old Stanton said the choice of Jewell was a no-brainer: "There was no one else, and I can't go on doing this forever. There's only one way to get a diver like that experience, and this was it."

Within hours they were bound for Thailand with mounds of gear—including a resupply for Stanton and Vollanthen. The tiny clique of the world's best cave divers carried their own customized kits, many of them homemade or produced in small shops.* This explains why Jason Mallinson drives around the UK and Europe in a van stuffed with his homemade or customized equipment. There are side mounts, underwater scooters, tubes to carry materials, and his own tanks. In an underwater world where you have only yourself to rely on, Mallinson trusts his own jury-rigged gear.

---

* Small manufacturers give their inventions creative names. Take, for instance, Doctor Duncan's Decidedly Dodgy Diving Device—or D6—a chest-mounted rebreather.

Because the American and British teams had ruled out the resupply mission to stock the cave with enough food and supplies to last the four-month monsoon season, the international teams had come to believe that they now had a small window of opportunity to dive the boys out. The international teams didn't know if diving out the kids would ever be approved by the Thai government, especially since they'd been unable to get the attention of the Thai leadership to brief them about the existence of the plan. But because preparations take time, they started to prepare as if the mission was a go.

In addition to packing some more food for the boys and their coach, they needed to get them diving supplies. Mae Sai is not a diving hub. The nearest dive shop is in Chiang Mai, more than four hours away by car. Many of the divers at the cave had brought extra wet suits, but they were for grown men, not emaciated boys. A floppy wet suit is worse than no wet suit. But the Chiang Mai Rock Climbing Adventures team contacted a company that manufactures wet suits for the apparel company Reef. They made a special order of kids' wet suits and had a driver deliver them immediately to the cave.

Under the ubiquitous blue tarps, Anderson's soldiers, Mallinson, and Jewell all got to work packing the

214 • MATT GUTMAN

wet suits and a few extra MREs they had scrounged up, tossing the castoff napkins, heating kits, and tiny Tabasco bottles into a growing pile of trash.

Mallinson set off first from Chamber Three. Jewell followed a few minutes later. Like their compatriots, the British duo of Jewell and Mallinson were accustomed to these wretched conditions. The sumps they explored in northern England were equally murky and even colder. Even so, Jewell and Mallinson found the going tough.

"The actual speed of the water was of a quite high velocity, and I'd be fighting against the flow in several sections," recalls Jewell. But since there were no other divers beyond Chamber Three, Jewell considered the visibility "quite good." For him that meant the ability to see three feet ahead of him, or about as far as his outstretched arm. It was enough for Mallinson and Jewell to start building a mental map of the route they would be taking several more times.

As if fighting the current weren't enough, the water itself was foul. Urine was being dumped into it from both Chamber Three and Chamber Nine—where feces was now being added into the mix. They were also concerned that there was runoff from animal farms up above, namely fertilizer and pig excrement, so they tried not to let too much water into their mouths.* But all divers end up ingesting some water. No one who worked inside the cave would end up leaving without a microbial memento.

Mallinson surfaced in Chamber Nine first. Before he saw anything, he could smell that the seventeen people in the tomb had already tucked into the MREs; the smell of highly processed American food was now added to the aroma of human waste and sweat.

Mallinson began pulling himself up the steep bank and detaching his gear. In his kit was another confined-space air-quality monitor—the one the USAF Special Tactics team had flown in on the C-130. He and Jewell had been tasked with taking another reading. He didn't need to check it. He began panting as he crawled up the bank. The air was thin, and it reeked. It was hard to breathe.

"We had spent a lot of time in caves and other places with bad air quality, so we quickly determined that the air had deteriorated," Mallinson recalled. There was no circulation in the chamber, so whatever air was there had been entombed with the boys when the water

---

* The amount of runoff produced by the few tiny communities atop the mountain was likely negligible.

closed off any exit. It wasn't only that oxygen was decreasing but also that carbon dioxide, the by-product of breathing, was increasing. Carbon dioxide is like oil in a salad dressing. It's heavier than oxygen, so if left unstirred it sinks to the lower parts of the chamber, building up exactly where everyone is trying to breathe—it can be especially dangerous when a group is sleeping.

Rising to his feet, Mallinson saw thirteen spectral figures hanging back in the darker recesses of the chamber. One or two of the boys stirred and came down. The boys looked like little spacemen wrapped in their foil blankets, stick legs poking out from below. One of the Thai Navy SEALs had stripped down. He was in his underwear, wrapped in a foil blanket—he'd given all his clothes to the boys. For his suit of underwear and silver "cape," he earned a nickname from the boys: Superman.

Mallinson introduced himself, and Dr. Bhak served as translator. The children, he said, "were pretty chilled out"—hardly panicky. They "understood the situation they were in and they were just hopeful that we could help them out."

Mallinson pulled up his tubes and started unloading the food and the wet suits. With Dr. Bhak translating, Mallinson, who is as chatty as a brick wall, did his best to work up a rapport with the boys. He asked if they were okay. "Okay, okay," Adul and the others responded. He got out his oxygen meter and began preparing a fingertip pulse oximeter and blood oxygen sensor—the kind many primary care doctors use to quickly determine oxygen saturation in the blood and a patient's pulse. Most of the boys' oxygen levels were above 94 percent—which wasn't great, but was hardly terrible given their circumstances. Slightly more vexing were their heart rates. Some of the pulses he measured pounded away, others were more sluggish. The air-quality meter beeped 17 percent oxygen in the room. But having been in confined spaces many times, Mallinson wondered, much as Stanton had during his reading, if the meter had been properly calibrated. After thirty years of diving and rescue, his gut feeling was that if the air quality dipped much further the boys would be in grave danger.

Jewell surfaced as Mallinson was measuring the air. He knew he was in the right place because he could hear the beeping of the air-quality monitor. He swam to the ramp, noted the thickness of the smell and the thinness of the air, and clipped his cylinders to a rope that Stanton and Vollanthen had rigged on their second dive.

The boys' bodies had been stripped of muscle and

fat. Their lungs strained for air. Some exhibited symptoms of pneumonia, including hacking, malaise, and shortness of breath. But, says Jewell, "I was very impressed with their state of mind. They took everything in their stride and they never showed any signs of doubt or upset or discomfort. Very brave boys."

That was essential, because he and Jewell had some news to deliver. They needed to have a frank talk with the boys. Through Dr. Bhak they explained that they had two options. They could stay in Chamber Nine for three to four months until the rains stopped and the water subsided. At that point they might be able to walk out. Or, Mallinson told them, "There is a possibility that the expert divers could dive you out."

Mallinson and Jewell knew the boys had a weighty decision to make. The boys likely didn't want to stay in this stone prison for a minute longer. But being dragged through a treacherous tunnel of roiling black water was also unappealing. The divers told them they should discuss it among themselves, think it over for the night. The civilian Australian team of divers headed by Dr. Harris would be coming in the next day to conduct a more thorough medical checkup and assess their fitness for a possible dive. The boys could inform them of their decision.

They turned to leave. But suddenly Mallinson had an idea. A Thai Navy SEAL major had given him a waterproof pad and pen containing a set of written orders for the SEALs. Mallinson had also used it to jot down the boys' medical readings. And the thought struck him: *These boys and their parents haven't communicated in two weeks. It's not a telephone line, but it's better than nothing.*

"It was quite impromptu. So I passed the pad to each of the kids and said, 'You've got half a page there, write a message to your parents.'"

Mallinson saw their eyes widen with delight. They spent a few minutes jotting down two or three lines each to their parents. "I think they wanted to put their parents' minds at rest. They wanted to say, you know, I'm in a situation, but I'm doing okay. Don't worry about me too much," said Mallinson.

Watching the boys scribble away, he was struck by a depressing thought: these could be the very last messages the boys sent to their families. Their very last communication.

Mallinson, the veteran rescuer, knew that the boys were one torrential downpour away from being marooned for weeks, possibly months. They had enough food to last them a week, maybe two. How long they could subsist on that fetid air was a complete mystery. "So it was quite emotional for me to be able to pass those messages back," said the man who had warned me at the start of our interview that he excelled at detail but wasn't so good at emotion.

Mallinson and Jewell packed their gear and headed back into the gloom. It would be hours before they arrived back at camp, bearing what Mallinson expected might be the boys' last communication to their families.

What Mallinson did not expect was that within hours of his return the letters would be published—and that suddenly every news outlet carried them verbatim:

Eleven-year-old Titan wrote:

Mom, Dad,
   Don't worry, I'm OK, please tell Yod to prepare to take me to eat fried chicken.

                                                    Love you

Thirteen-year-old Dom wrote:

I'm fine, but the weather is quite cold. But don't worry. But don't forget my birthday. (Which had been on July 3.)

Fourteen-year-old Adul wrote:

Now, don't worry about us anymore. I miss every-body. I really need to go back home.

Sixteen-year-old Night wrote:

Night loves Dad and Mom. Don't worry about Night. Night loves everybody. (Beneath the bubbly Thai writing he signed off with a pair of hearts and his name.)

Thirteen-year-old Mark wrote:

Mum, are you at home, how are you? I'm fine. Can you tell my teacher." (He was apparently worried about his upcoming exams.)

Fourteen-year-old Biw wrote:

Don't worry, Dad, Mom, Biw has just disappeared for only two weeks, I will go back and help Mom at the store as soon as I have a free day. I will rush to go back.

And Coach Ek's letter to the parents was particularly poignant:

> *All the kids are fine. There are people taking really good care of them. I promise I will take care of the children the best I can. Thank you for your support. I'm really sorry to the parents.*

A few days after the Sleeping Princess swallowed up the soccer team, a midlevel Thai monk had gone to confer with the cave spirits. According to Thai officials who met with him, he came away shaken. The monk told them that the spirits of the cave demanded the sacrifice of a cow, a buffalo, and two men. However, even the more superstitious rescuers didn't pay much attention to the prophecy until the morning of Friday, July 6.

As the British diving team worked on getting supplies to the boys, the Thai Navy SEALs were still working on their oxygen hose project. Petty Officer First Class Saman Gunan, the square-jawed ex-SEAL who was photographed in some of the higher-level planning meetings, had been working at the site for days; because of his smattering of English and can-do attitude, he'd become acquainted with foreign rescuers like Stanton, Vollanthen, and Vern. Late on July 5 and into the early morning of July 6, Gunan was ferrying tanks in the sump between Chambers Two and Three. It was a short dive, but he had apparently gone back and forth a number of times and was already exhausted. It was the end of his shift and—according to the Thai SEALs—his last dive of the day. Gunan was a chiseled triathlete. He was arguably one of the most fit men taking part in the rescue. But something happened on his way from Chamber Two to Chamber Three. For unknown reasons, Gunan's mouthpiece and mask popped off, according to Rear Admiral Apakorn.

The SEALs say he was carrying three tanks with three regulators, but they were apparently floating octopuslike in the water—making them more likely to snare on the loose wires, stalactites, and hoses littering that sump. His dive buddy later said that through the murk, it looked like Gunan was trying to grab one of them, but he couldn't find it, or perhaps his hands had become snagged on something. His dive buddy started kicking frantically toward him. By the time he reached him, Gunan had gone limp.

His dive partner tried to plug his own regulator into Gunan's mouth, but he was already unresponsive. His dive buddy dragged him back to Chamber Three. As soon as he reached the air, he spat out his regulator and screamed for medics. The Thai SEALs had hoped to resuscitate him, as they had the other foundering SEAL only a day earlier. But Gunan wasn't breathing. They tried to scramble a quick rescue, but it would take well over an hour to get him out of the cave. He died somewhere along the way.

While Rear Admiral Apakorn has spoken about the incident, the Thai Navy SEALs have not released a comprehensive report on Gunan's death. Some of the foreign divers quietly blamed the ill-fated hose project or wondered if he had been given an empty tank by accident. They also wondered whether he had been inadvertently poisoned. International divers told me that somehow during one of the support team's nightly refills of the camp's air tanks, carbon monoxide from the compressors was mixed with the air in some of the tanks. That might help explain how a spectacularly fit man had died on a thirty-foot dive—a swim about as long as two Volkswagen Beetles parked end to end. It was a stark reminder of just how high the stakes were.

The death sent jitters through certain parts of the Thai rescue community. At the hotel where the ABC team was staying, the head of communications for the cave site told me they were forging ahead anyway, but his men were spooked. Some were now tormented by the midlevel monk's prophecy a day earlier, predicting the cave would claim "two men" in exchange for the soccer team. They feared that another rescuer would die.

The next afternoon, Saturday, July 7, I watched a pair of white Toyota minivans start weaving around the workers and journalists walking up the hill from the checkpoint on the main road below to the campsite outside the cave. When the two vans stopped, a crowd of Thai military and police officials began excitedly crowding around them. The doors slid open and out stepped a few monks with shaved heads, their bright-saffron robes covering one shoulder. Men in reflective vests and a few more monks then crowded the door of the main van as a fleshier, older monk with big sprouts of white eyebrows was eased out. It was Kruba Sangla, a monk elder who was much revered.

The Thai press sprinted toward him, snapping pictures. Some immediately began live-streaming the event. Sangla was there to right the earlier prophecy by the upstart monk predicting two deaths.

He was helped up the stairs, past the security barrier leading to the main operations center, and up the hill toward the mouth of the cave. He had some business to attend to—a negotiation with the cave spirits. According to the Thai communications manager, who showed me pictures of the ceremony, the senior monk began bargaining with the cave spirits. Instead of the buffalo, the cow, and the two men, he would offer a wild boar, a white rabbit, and thirteen chickens—symbolizing the twelve members of the soccer team and their coach. The menagerie offering was placed in the jungle near the cave; the animals were tethered to a bench and left alive.

Chapter Thirteen
The Wet Mules

They started tearing into the MREs. Apparently without consulting the rescue divers, Dr. Bhak and the three SEALs with him decided that instead of the allotted one meal a day, everyone would get three. It was, after all, his mission to dose the boys with as much care and with as many calories as they could consume. Whatever lay ahead, they needed to regain their strength after twelve days of hunger and muscle atrophy. Besides, the boys didn't give him much choice. As soon as they finished one meal pack they demanded another. The boys actually liked them. Typically they would get one main course, then a starch—like a muffin top or rice—and then a dessert. And yet nothing could fill them up; they craved more.

So for the first two days, as they crouched in the dark, Dr. Bhak put the boys on the spelunking equivalent of bed rest. They didn't talk much because their main activity was sleeping, which was a good thing for Dr. Bhak. Because of what the British divers had innocently said when they found them a few days earlier, the boys and their coach originally expected the SEALs to promptly deliver them from the cave. After all, the Brits had told Adul, "Tomorrow, we'll come with an ambulance." The Thai SEALs had to explain that rescue might take some time and they should trust that the authorities were doing everything possible to get them out. Dr. Bhak was likely unaware of the monsoon systems grinding toward them. Should those rains arrive before the rescue mission, he might have unknowingly complicated the group's survivability by doling out multiple MREs instead of one—however well-intentioned his actions and however hungry the kids were.

By Dr. Bhak's third day there, Friday, July 6, the boys started perking up. There was a lot of talk about where they would travel when they got out and what they would eat. They desperately wanted to be out before the soccer World Cup finals in the middle of the month. Dr. Bhak remembers this period as rather monotonous: "Our usual routines were waking up, eating our meals together, talking to each other for a bit, and then sleeping."

All that eating jump-started their digestive tracts. With seventeen people now consuming about twenty-five hundred calories per day, defecation became an issue. They had dug a latrine where their little mound dipped down toward the cave wall. When it filled up with urine or stank too much, they'd cover it. But defecation required some more creativity. So the boys, always with a SEAL escort, would paddle about ten feet out into the canal and float until the deed was done. The current would carry their waste back toward the T-junction—in the direction of incoming divers. It was through these bowel-movement excursions that the Thai SEALs learned that all the boys could swim, more or less. MREs are surprisingly tasty (possibly due to their high sodium content), and with their energy up the boys' internal thermostats had stabilized. They were no longer as cold and didn't mind going into the water.

But there is an ocean of difference between being able to paddle out ten feet and being ready to embark upon a mile-long underwater odyssey, in a full face mask, in molasses-colored water . . . as your first scuba dive ever.

At press conferences outside the cave, government

officials made it sound like the boys were in a professional diving course. Even before Mallinson and Jewell had returned from their delivery run of food and wet suits to the boys, Deputy Prime Minister Prawit Wongsuwan told reporters in Thailand, "The [current] is very strong and space is narrow. Extracting the children [will require] a lot of people. . . . Now we are teaching the children to swim and dive." This is the same deputy prime minister who told reporters that the U.S. Special Tactics team possessed special mountain-penetrating radar.

In reality, there was no way that the boys would be swimming out, diving out, or in any way actively participating in a possible rescue. Everyone who had any involvement with the planning of the emerging rescue operation—which at this point the Thais considered the least desirable option—considered it a nonstarter. Between the oxygen levels, the weather forecast, and the inability to adequately resupply food to Chamber Nine, the situation had undoubtedly become more dire. Almost everyone on the international teams understood the new stakes involved in this operation, but somehow—possibly owing to a messenger's reluctance to deliver bad news—this reality had apparently not yet made its way up to the Thai decision makers. Which is surprising given that Vern Unsworth, who knew the cave better than anyone, had been trying to tell the Thai leadership since June 25 that "this was a rescue mission of very high risk. Very high. But I told them if you wait until December or January you'll be bringing out thirteen bodies." "High risk" had become a euphemism among the divers for "we expect many of the boys to die." Vern says his calls for action caused him to fall out of favor with the leadership. They didn't want to hear it, he said.

The apparent disconnect persisted despite the fact that the main war room and the moldy British diving office were next door to each other. Yet somehow the Brits' calls to launch an urgent rescue operation had not gotten through. The Thai government seemed unaware of the worsening conditions in the cave and the necessity of prompt action, opting instead to keep pushing for unworkable solutions like the oxygen tube, drilling, the boys swimming out themselves, and—arguably the most dangerous of all—extreme patience.

Such was the state of things when Richard Harris and Craig Challen arrived at the cave complex on July 6. The two middle-aged Aussies belong to a cave-diving group calling itself the Wet Mules. One of its members, who traveled frequently to the United States, had come across the expression that a wealthy man "has enough money to burn a wet mule." The colorful phrase stuck.

As their tongue-in-cheek Web site says, "it began to enter our conversations frequently . . . hungry enough to eat a wet mule, as tired as a wet mule, as wet as a wet mule, etc. Then it occurred to us! As a large part of our chosen pursuit of cave diving seems to revolve around ferrying heavy objects in and out of caves, submersing ourselves in frigid waters for many hours and generally abusing our bodies in a multitude of ways, we were beginning to take on the persona of the wet mule itself!"

The Wet Mules would stalk the Nullarbor Plain, off Eyre Highway, which boasts arguably the world's longest straightaway—ninety-one miles of dizzyingly unvarying two-lane blacktop. It's so flat today because about 100 million years ago the Australian landmass was split and its central belly was covered by a blanket of ocean. It's now the world's largest single exposure of limestone bedrock. When the ocean receded, rainwater honeycombed the porous limestone, creating one of the earth's greatest (still largely unexplored) cave systems.

It beckoned to cavers who were, as the Mules called themselves, "Stubborn, strong of back and oblivious to pain."

"We love," their amusing Web site continues, "unexplored caves; making it up as we go along; combining diving with helicopters; unique solutions to unique problems; rebreathers, SCUBA cylinders, snorkels or whatever will get the job done safely and efficiently; individualism; contributing to the science, conservation and understanding of what we enjoy . . . caves!"★

Over the years they spent most of their free time exploring them. Among their great stalwarts were Challen and Harris. Challen, the veterinarian, had recently been named Australasian Technical Diver of the Year by his country's biggest diving trade show organizer, Oztek. He had set an Australian depth record of 722 feet (nearly as deep as the Eiffel Tower is tall—at that depth your lungs are squeezed to about one-twentieth of their size at the surface) in a place called the Pearse Resurgence in New Zealand. The resurgence is the artesian-spring headwaters of the Pearse River; it plunges to a still-unknown depth below the Kahurangi National Park. With its wind-scrubbed cliffs, alpine pastures, and temperate rain forest near the river, the park is a place where divers from around the world congregate—including Stanton and Vollanthen. In fact, Stanton teamed up with Harris in 2007 on one of his expeditions there. In a

---

★ For an entertaining read and a good primer on Australian cave diving and Wet Mule antics, check it out for yourself: www .wetmules.com.

magazine article he later wrote about the experience, Harris describes nearly dying when he became dizzy and disoriented and found himself trapped in a subterranean cavern; he only managed to escape when fellow divers came for him.

If Challen had scraped below the radar, Harris seemed an ever-present blip. He has published widely—magazine articles about the serenity of the deep and its dangers. He has described descending into the crystalline waters of one of the caves of the Nullarbor Plain as "being in space," where the walls are so white that lights reflecting off them magnify the beauty of the water and the visibility is hundreds of yards. It was a nearby cave that claimed scientist and fellow cave diver Agnes Milowka. She became disoriented in one of the passages, ran out of air, and died. Harris was the diver who found and helped recover her body.

Both Milowka and Harris had consulted and stunt-dived for James Cameron's 3-D disaster film *Sanctum*, loosely based on an ill-fated expedition to Nullarbor in 1988. A team of fifteen divers, led by veteran Australian caver Andrew Wight, had plunged in about five hundred yards when it started to pour. A freak storm pummeled the area with two years' worth of rain. Wight and a teammate managed to scamper out as boulders rained down and closed the once-gaping cave mouth. He then helped spearhead the twenty-seven-hour rescue effort to save the rest of his team. The screenplay of *Sanctum,* based on this experience, was cowritten by Wight.

That scenario wasn't so dissimilar from the rains that choked off Tham Luang: the main difference is that while the exact timing of northern Thailand's rains is unpredictable—like the storm that clobbered Nullarbor—Tham Luang's annual submergence is something you can bet on.

Even though there was no doubt that Harris's background as an anesthesiologist made him uniquely qualified to help in this situation, there was no template for sedating humans and then ferrying them through a mile of submerged cavern. The closest thing to a precedent anyone could think of involved animals. Research had been conducted on the sedation of wild pinnipeds (seals, sea lions, elephant seals, etc.) off the coast of California. The animals had been shot with sedative darts in order for scientists to temporarily capture the feisty creatures and conduct blood tests and ultrasounds. Ketamine, a common animal tranquilizer and once the gold standard of anesthetics in pediatric medicine, was most commonly used. After being darted, the animals would bolt into the water. The scientists noticed that even when the animals were unconscious in the water, they maintained their diving reflex—which prevented them from gulping seawater into their lungs.

The divers and doctors at Tham Luang hoped that the use of ketamine on the boys might work as it had with the pinnipeds, and that perhaps if the boys' masks flooded their own diving reflexes would kick in. But again, this was only a theory. It had never been properly tested by researchers on humans. It was an enormous leap of logic, and the team knew it was a gamble. Pinnipeds are what is called obligate nasal breathers, meaning they breathe only through their noses, and have been evolutionarily perfected for prolonged dives and breath-holding. Millions of years of adaptation to underwater survival have endowed them with a much higher tolerance for carbon dioxide in their systems—the buildup of which the brain detects when it sends signals to the diaphragm to contract, thus pulling fresh air into the lungs. Humans are not built for prolonged submarine exposure and have a far lower tolerance for carbon dioxide, and when we gasp for air, we do it through our mouths.

If there are any pleasant ways to die, drowning is not one of them. The carbon dioxide in your body triggers those contractions of the diaphragm, and a sensation of being choked overwhelms you, as if death itself is gripping your throat and clutching your lungs. Some feel a stabbing sensation in their eyes and ears, and then there's the involuntary opening of the mouth—maybe to scream, maybe to breathe air that turns out to be water. Once the lungs fill, your buoyancy is lost and you sink. Death follows quickly and mercifully.

In recent decades free divers have expanded our understanding of the human body's tolerance for the immense pressures of the deep. On a single breath, free divers have reached depths of over seven hundred feet—nearly as far as Challen dove on rebreathers. Lungs shrink to one-twentieth of their normal size and something amazing happens: the mammalian dive reflex kicks in. Heart rates slow by over 25 percent, and even as blood rushes to the internal organs to keep the diver alive, they typically don't lose consciousness even when their heart rates slow to about fifteen beats a minute (most people's hearts beat between sixty and eighty times per minute). But this happens at great depths, among divers who have trained to control their breathing and heart rates as well as yogis. It's not something the boys could have done—besides, their challenge wasn't depth, it was horizontal distance.

In the end, the concept of sedation remained controversial. While the arrival of Harris and Challen at the cave had been okayed by the Thai government, there was no definite plan for them. In fact, no one even knew whether the Thai government would allow a rescue operation to occur in the first place, let alone one that involved sedating thirteen children. Harris and Challen had made the four-thousand-mile trip from Australia to Thailand in the hope that they'd be able to do some good in the next few days, but no one knew if they'd even be allowed to suit up for the mission.

Even as their plan continued to coalesce—as Thursday, July 5, slipped into Friday, July 6—the international teams' sense of dread grew. The monsoons were only a weather system away, and they couldn't get the Thai SEALs to focus on anything but their oxygen hose. The USAF Special Tactics team and the British, in particular, felt they had to set the rescue in motion, and they had to do it immediately—but hadn't been able to pitch their plan to a decision maker. With the arrival of Harris and Challen, they now had six total divers (four Brits and two Aussies) with the chops to haul inert 70- to 150-pound sacks of human beings through fifteen hundred yards of underwater terror. Harris was to be the main doctor back in Chamber Nine. Challen was to provide medical support along the way. That left four primary rescue divers for thirteen people. The four

Thai Navy SEALs who had been with the kids would have to dive out on their own power; it was decided they would neither assist in the rescue nor receive assistance from rescuers on their way out. During their initial planning session, the Brits and the U.S. Special Ops team had already ruled out extracting all the boys in a single day. The divers expected it would take them about eight to ten hours to swim to the boys and ferry them back out. They said they would need about sixteen hours downtime after each rescue. At that pace it could take four days.

Hodges and his team started to get nervous. They knew major rain was expected after the weekend, as much or more than the rains that had swamped the cave a week before and had triggered the all-out evacuation and suspension of all dive activity.

In their tent, Hodges relayed his concerns not only to his team but up the command chain as well: "If we don't get this plan briefed and approved and everybody on board, we're probably not gonna be able to execute it, especially if we're gonna try to start doing it Monday or Tuesday. This is gonna be a huge effort. It's going to require a lot of prestaging in between each day and relocating equipment."

Outside the cave on July 6, Hodges and Anderson were lobbying for action. While the plan still wasn't firmly drawn up, they had a pretty good sketch of what the Rube Goldberg rescue would look like. What they lacked were supporters. That day they spoke to the Thai SEALs, recalled Hodges: "Hey—this piping in oxygen is absolutely your guys' mission. If that's what you wanna do, then we understand. But we don't think that that's a long-term solution." As it turned out, the SEALs had come to a similar assessment: after Saman Gunan's drowning, they quietly wrapped up their plan to lay an oxygen pipe through the cave; it was too complicated and clearly too risky.

The Americans knew they were days from being forced to maroon the children to a possibly gruesome death, and while they believed the SEALs also grasped the gravity of the situation, that message had apparently not been communicated to the military's decision makers. In the U.S. military, a commanding general would typically have the authority to make the tactical decisions necessary to execute the mission he'd been assigned. But the tactical decisions determining almost every facet of the cave rescue in little Mae Sai were being made by the political hierarchy in Bangkok. It was by no means ideal.

"Was there frustration? Absolutely," said Anderson. "I don't think we were frustrated with any one individual. We were frustrated by the process of having to go up to the highest, highest levels. And I also think that there was a disconnect between what they thought the severity of the situation was at the highest levels and what was actually happening down at the tactical level."

Approval or not, the international team had to keep pushing its preparations, and by July 6, they were well under way. A day earlier, the Americans had assigned the so-called Euro-divers—Claus Rasmussen, Ivan Karadzic, Mikko Paasi, Nick Vollmar, and Erik Brown—to help with provisioning the route and tidying it up. The Euro-divers' first mission that day was to find a safe spot somewhere before the T-junction to stash fresh air tanks. Basically, they were to set up a gas station at a midpoint of their choosing along the waterway between Chambers Three and Nine. They would leave fresh tanks and diving equipment such as regulators for the Brits and Aussies shuttling back and forth. It wasn't particularly glamorous work; it wasn't particularly easy, either.

They had been tasked with taking three extra tanks each, to amass a cache of twelve. The packs of three air tanks they were to deliver deeper into the cave were waiting for them at Chamber Three. Surprisingly, given all they'd heard about the conditions in the water, they coasted at first. The Euro-divers had been told by

Ben Reymenants to leave their fins at home. The current was so strong they wouldn't need them—they'd be using their hands to crawl and pull their way through. Now the current had eased and the going was smooth. They devised a system in which they lashed their tanks into a three-pack, wedged with Styrofoam to prevent the tanks from dragging on the bottom. On solid ground they weighed about a hundred pounds; in the water it felt like they weighed only a couple of pounds.

Claus Rasmussen, the other Dane on the team and the unofficial leader, headed in first. He had been confused by their orders. Rasmussen recalls at least three distinctly different sets of orders that day, a couple of which had him heading in first to clear debris, telephone and electrical lines, and remnants of the ill-fated oxygen tube that had been abandoned. The Thais at the sump in Chamber Three told him not to bother, but he figured he might as well take a peek and try it anyway.

He wormed his way past the sump at Chamber Three and worked through the tangle of lines. He dropped down low toward the gravel at the cave bottom to perform his housekeeping assignment, winding the loose wires around his forearm, cutting off sections, and then discarding them in nooks behind rocks. As he was working he could feel other divers come past. The first one to glide by he knew to be Paasi. Then another fig- ure passed, closer to him. Rasmussen felt his own hand jerk in the direction of the diver. He could feel the cable on his arm start to unspool as the diver swam past. He immediately let go, hoping that whatever it was getting tangled around was not the metal part of the diver's rebreather. As soon as the wire was out of his hands he swam after the diver to warn him. It didn't take long to reach the diver—he could see that it was Vollmar and could hear that the wire had become tangled around his breathing apparatus. Vollmar switched to his backup scuba tank with an open-circuit regulator (the kind that lets out bubbles when one breathes instead of re- cycling them as a rebreather does).

Hovering in the tunnel, Vollmar was fiddling with his rebreather and trying to untangle the wire—but it was hopelessly ensnared. He wrestled with it for a few minutes, then made a decision and started unclipping the three tanks he was transporting and resetting them on the guideline. He was bailing out. He had plenty of air in his extra tank to get back, but clearly not enough to complete the mission—especially while encumbered with a broken rebreather. Rasmussen gave him the okay sign and clipped Vollmar's three tanks onto his own harness. Later that day Vollmar spent a couple of hours tinkering with his gear, but it was wrecked and he was out for the rest of the rescue. Shaking his head at the memory, Rasmussen said that if it had happened deeper in the cave it might have been disastrous.

Paasi had been at the head of the small procession of four divers, oblivious to the near disaster in the line behind him. At the mouth of the cave he'd been given instructions; after the tight restriction past Chamber Three—a 150-yard-long crawl space littered with electrical cables from the first few days of the dive—he would come to a junction at the start of Chamber Four. To him it looked like a fork in the road, though he didn't know that both routes eventually would converge. He had the option of taking a left or a right. He was told to take a right. When he arrived at the junction, he decided to wait for the other divers to catch up. But minute by minute he grew more disoriented. He didn't know that Vollmar's rebreather malfunction had held up the line. Floating in the dark, the minutes dragging on, Paasi looked at his hands and noticed that even with his headlights shining directly on them the tips of his fingers seemed to dissolve into the brown plasma of the water. Paasi kept a thumb pointed at the exit, but "the dark plays tricks on your mind." This was a section of cave that would lead to another potentially fatal incident four days later, when it would matter most.

The "Wild Boars" soccer team bicycle group. Coach Eakapol is at the far left wearing helmet and sunglasses. FROM LEFT TO RIGHT: Eakapol Jantawong (EK) / 24; Chanin Viboonrungruang (TITAN) /11; Prachak Sutham (NOTE) / 14; Phiphat Phothi (NICK) / 15; Phonchai Khamlu-ang (TEE) / 16; Somphong Jaiwong (PONG) 13; Phanumas Saengdee (MICK) / 12; Natthawut Thakhamsong (Tern) / 14; Adul Samon (ADUL) / 14; Mongkhon Boonpiam (MARK) / 13. *Seated left to right:* Duangphet Phromthep (DOM) / 13; Ekkarat Wongsukchan (BIW) / 13. (*Courtesy of the Thai government*)

The mouth of the Tham Luang cave in Mae Sai Thailand. Its first room can fit a Boeing 747. (*Courtesy of Petpom Tolmuang*)

An image taken from the back end of Chamber One looking toward the mouth of the cave. The flooding has begun. From July through December or January the entirety of the cave is flooded. (*Courtesy of Vernon Unsworth*)

Mae Sai Prasitsart School. Twenty-eight hundred students attend, including at least six members of the Wild Boars. *(Courtesy of the author)*

Tham Luang Park Ranger discovers one of the boys' soccer bags. Inside are some phones, clothes, and Wild Boars jerseys. *(Courtesy of Petpom Tolmuang)*

Pumping operation at Tham Tsai Tong. Engineers lowered the water table of the entire region in an effort to expedite the draining of floodwaters from the cave. *(Courtesy of Thanet Natisri)*

The author interviews Governor Naronsak Osatanakorn, the Incident Commander of what would soon become a massive search and rescue operation for the boys. *(Courtesy of ABC News/Robert Zepeda)*

British caver Vernon Unsworth briefing leading Thai officials including Interior Minister Anupong Poachinda on June 26. It was at this meeting that he warned officials a rescue dive would be necessary to save the boys. (*Courtesy of Vernon Unsworth*)

From left to right: Caver Rob Harper and rescue divers John Vollanthen and Rick Stanton. (*Courtesy of Vernon Unsworth*)

A submerged section of the cave between Chambers One and Two. (*Courtesy of the U.S. Air Force*)

The muddy, hose-crowded steps leading up to the mouth of the cave. In the green hut at the top right corner of the picture is one of the Spirit Temples of the Sleeping Princess. (*Courtesy of Asaf Zmirly*)

USAF 353 Special Operations Group Major Charles Hodges shaking hands with Maj Gen Bancha Duriyaphan, commander of the 37th Military District who led efforts to find alternate ways of extracting the boys, such as drilling an escape shaft or discovering an unknown entrance to the cave. (*Courtesy of the U.S. Air Force/ Captain Jessica Tait*)

Thai troops carrying supplies to the mouth of the cave, past a giant mobile generator. (*Courtesy of Asaf Zmirly*)

USAF Special Tactics operator at a helicopter landing zone being cleared from the jungle above the cave. It was part of an unsuccessful effort to drill rescue tunnels to the boys. (*Courtesy of the U.S. Air Force*)

USAF Special Tactics operator Master Sergeant Derek Anderson (with pen) and Major Hodges, going over a hand-drawn map of the cave. (*Courtesy of the U.S. Air Force/Captain Jessica Tait*)

Thanet Natisri explaining his water drilling efforts to Colonel Singhanat Losuya of the Thai 37th Military District based out of Chiang Rai. Col. Losuya would later play a critical role in advocating for a rescue dive. (*Courtesy of Thanet Natisri*)

Members of the Chiang Mai climbing team taking a break from scoping out possible alternative entrances to the cave. From right to left, Josh Morris is third in. Mario Wild (glasses and hat) is eighth in from right to left. (*Courtesy of Chiang Mai Rock Climbing Adventures—CRMCA*)

USAF Special Tactics Captain Mitch Torrel speaks to members of the Thai Navy. Torrel was ultimately tasked with commanding U.S. operations in Chamber Three, including removing the boys from the water and assessing their medical condition. (*Courtesy of the U.S. Air Force/Captain Jessica Tait*)

Ben Reymenants pointing at a map of the cave route with Australian and American troops, including USAF Special Tactics captain Mitch Torrel, looking on. (*Courtesy of the U.S. Air Force/Captain Jessica Tait*)

The only way to ferry equipment into the cave is on the shoulders of soldiers or rescue workers. It would typically take rescuers carrying gear ninety minutes to walk, crawl and wade from the mouth of the cave to the end of Chamber Three. (*Courtesy of the U.S. Air Force*)

Monk Boonchum Nanasamavara, of the "Thai Forest Tradition," is particularly revered in the north of Thailand, parts of Laos, and Burma. The monk (with maroon head covering) and his followers walk barefoot in the ankle-deep mud. He prophesied that all of the boys would be found and rescued. (*Courtesy of Asaf Zmirly*)

Petty Officer Saman Gunan. The ex–Navy Seal died during a diving operation in the cave late on July 5. (*Courtesy of Noppera Bosri*)

The author reporting for ABC News outside of the cave. (*Courtesy of ABC News/Chris Geerdes*)

On the night of July 2, after Vollanthen and Stanton found the boys in Chamber Nine, the Seals, the Brits, European divers and American representatives met to discuss rescue options. At that point Vollanthen and Stanton considered a rescue dive impossible. (*Courtesy of the U.S. Air Force*)

Stanton's girlfriend, Amp, and Stanton prepare a map for a presentation with Thai leaders. (*Courtesy of Rick Stanton*)

USAF Sergeant Major Derek Anderson briefing Thailand's Interior Minister, members of the Royal Guards, Thai military leaders, Governor Narongsak, and others about a possible rescue plan on the night of Friday, July 6, 2018. (*Courtesy of CMRCA*)

Hodges fits a young local boy with a positive pressure mask. Since the rescuers couldn't try the masks on the boys trapped in the cave, they had to test the masks on local boys of a similar age. (*Courtesy of the U.S. Air Force*)

Saturday, July 7, divers practice the rescue dives with local boys at a Mae Sai pool. Note the large dive tank attached to the boy's chest. (*Courtesy of the U.S. Air Force*)

Divers with scuba gear at Chamber Three. (*Courtesy of Asaf Zmirly*)

Rehearsal of Concept (ROC) Drill. Those water bottles represent air tanks, and the chairs represent chambers along the rescue route. Rescuers walked through multiple times on Saturday, July 7, to rehearse the choreography. (*Courtesy of the U.S. Air Force*)

Stanton preparing to dive food to the boys. The Thai officers with rocks in their hands are helping him make the tube neutrally buoyant by weighing it down. (*Courtesy of Rick Stanton*)

USAF Special Tactics Sergeant Sean Hopper briefs members of the Chiang Mai ropes and climbing team. (*Courtesy of CMRCA*)

Some of the hundreds of air tanks and compressors at camp. (*Courtesy of the U.S. Air Force*)

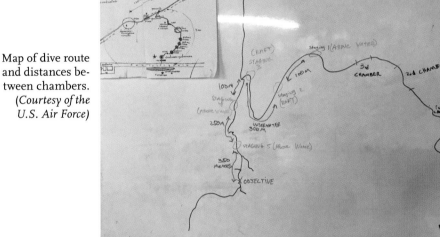

Map of dive route and distances between chambers. (*Courtesy of the U.S. Air Force*)

Thai diver at the bottom of the first chamber. Note: one of the very few benefits of all that mud was that it made it easier to keep air tanks upright. (*Courtesy of Asaf Zmirly*)

Rescuers in waist-high water near Chamber Two. (*Courtesy of the U.S. Air Force*)

Part of the elaborate rope systems set up in Chamber Two operated by USAF Special Tactics operators and members of the CMRCA. (*Courtesy of the U.S. Air Force*)

A boy being lowered on one of the rope systems in Chamber Two. This image shows the difficulty of the terrain, especially in this part of the boulder field in Chamber Two. (*Courtesy of the U.S. Air Force*)

One of the boys placed on a floating litter and taken over a partially flooded area. (*Courtesy of the U.S. Air Force*)

USAF Special Tactics operators carrying a boy in a Skedco through part of Chamber Two. Note the sharp decline on the right side of the picture. (*Courtesy of the U.S. Air Force*)

Rope system lowering a boy in a Skedco to a Thai SEAL. (*Courtesy of the U.S. Air Force*)

Close-up of a boy in a Skedco. Note the air tank filled with oxygen near his left leg. The positive pressure mask was left on the boys and coach during the entirety of the rescue. (*Courtesy of the U.S. Air Force*)

Thai rescuers bringing a Skedco down stairs near Chamber One. (*Courtesy of the U.S. Air Force*)

Author reporting as the first ambulance, carrying fifteen-year-old Note, leaves the cave site. (*Courtesy of ABC News/ Robert Zepeda*)

Thanet tinkering with Elon Musk's escape pod. While rescuers said it was impractical for the Thai cave rescue, Stanton believes it could be used in future rescues. (*Courtesy of Thanet Natisri*)

LEFT: The boys' families thank rescuers after the rescue on July 10, 2018. Josh Morris, with hand on heart, translates. (*Courtesy of CMRCA*) RIGHT: The author interviewing Master Sergeant Anderson at Kadena Air Force Base in Okinawa, Japan. (*Courtesy of ABC News /Robert Zepeda*)

Chiang Mai climbing team and rescue divers pose for a post-rescue picture. *Back row, from left to right:* Connor Roe, Jim Branchley, Chris Jewell, John Vollanthen, Craig Challen, Dr. Richard Harris, Sirachet "Add" Kongsingh (Chiang Mai climbing), Gary Mitchell (BCRC), Rick Stanton, Vernon Unsworth, Martin Ellis, Mike Clayton (BCRC), (on the far right in green shirt) Josh Morris. *Bottom Row, from left to right:* Adithep Khamsri (Thep), Jason Mallinson, Jim Warny, Nappadon Uppakham (Taw), Archan Nawakul (Toto), Jarundech Tongnak (Jojo) Mario Wild, and Corporal Phanlop Boonkham. *(Courtesy of CRMCA)*

Some of the boys at the Wat Doi Wao temple in the weeks after the rescue. After their rescue, the twelve boys became apprentice monks for nine days as a way to honor ex–Navy SEAL Saman Gunan. *(Courtesy of the author)*

In the dark, spinning around for a while, his three choices looked identical: forward to the boys, back to the exit, or the turn into a dead end. There was no way to orient himself, so he just clung to the rope. Finally the next diver came and pointed him in the right direction. It was his first lesson in the treacherousness of the cave.

Minutes later, Karadzic got stuck in a restriction. He eventually wiggled his way out, but he found the experience instructive. Most divers follow a line close to the bottom. Divers swim horizontally, their faces looking toward the floor and the guideline beneath them. Lifting one's neck to look forward or up is not only difficult, but is sometimes made impossible by the dive valves and hoses on their back-mounted tanks. But the guideline that had been set up in the cave was high and loose. Its slack would allow it to roam more than three feet in one direction or another in a tunnel, potentially stuffing a diver face-first into a restriction or unseen wall. It also required the divers to choreograph moving the trio of tanks they towed from one hand to the other, while ensuring the line lashed to the tanks didn't snag on the main guideline. It was like walking several unruly dogs on loose leashes.

The team managed to find their "gas station" before the T-junction, at nearly the exact geographical midpoint between the boys and the mouth of the cave, and started unloading their tanks. This would become known as Chamber Six, and while Karadzic and Erik Brown didn't it know it yet, they'd be spending a lot of time there. Over the next four days they would make as many trips as they could handle, ferrying in more than one hundred air tanks; in the course of these trips they became so familiar with the cave's turns, dips, and lacerating edges that they could have navigated it blindfolded. They were to prove invaluable stevedores: no one else in camp—other than the British foursome and the Aussie doctors—had the experience necessary to stockpile all those tanks.

After that first dive, the Euro-divers trooped out exhausted. One of the U.S. Special Ops team's captains, Mitch Torrel, came to Karadzic with a big blue tarp.

"Hey—anything you need to take in to Chamber Three, we'll haul in for you," said Torrel.

And for the next five days, the Special Ops team— elite soldiers trained at the cost of millions of dollars each—would crawl, pull, and carry hundreds of pounds of other people's gear into Chamber Three. It would be neatly set up for the divers when they arrived.

Now all they needed was an approved mission that would put those supplies to use.

Chapter Fourteen
Sticking Your Neck Out

They needed to get the Thai government on board. On Friday, July 6, getting the green light from the Thai government became the Americans' primary focus. The effort had begun with Anderson's conversation with the Thai Navy SEALs that day, but that conversation was only the beginning. Until then, the details of the proposed rescue operation had not been discussed seriously among the top leaders in camp. That was because the American team planning the operation had been unable to get an audience with the decision makers. They had to find a way to break through the bureaucracy, and in the end, this ultimately required a bit of serendipity in the form of an American rock climber named Josh Morris.

Morris had bailed on a trip to the United States

to volunteer at the cave site. He was a sinewy mountain climber who had lived in Thailand for seventeen years, married a local woman, had four kids with her, and possessed a voice that could easily have made him Nicolas Cage's voice double. He had founded Chiang Mai Rock Climbing Adventures about fifteen years earlier and played a significant role in developing rock climbing in Thailand. Before the boys went missing, he'd planned to go back stateside for a grand tour during the last week in June—a wedding, his father's eightieth birthday, a reunion with grade-school buddies. It didn't happen, because he had sent his Chiang Mai climbing team to volunteer at the cave site. After a few days he followed.

As the days leading up to the big family trip to the U.S. neared, he finally broke the news to his kids. His five-year-old daughter, Kamine, cried—she'd been looking forward to playing with her cousin Ariel, and had been preparing by repeatedly watching *The Little Mermaid*. And then Josh did what daddies do—he exaggerated: "These boys want to go see their mommies and daddies, too, and if you can help me support that, maybe we can make that happen." In truth, he didn't know if he'd play any role at all in the rescue.

By July 6, it had become pretty clear that there wasn't much reason for Morris to be there. Indeed, he'd been on site for a week or so and spent most of that time wandering around the camp and the hills looking for something worthwhile to do. Most of the efforts of Morris and the rope-climbing experts from his company—including Taw, who was among the first to the cave, and Mario Wild—had been devoted to finding an alternative entrance to the cave. While the divers had been discovering the kids and devising a plan to dive them out, Morris's group had continued helping the Thai military sniff out the mountain above the cave for shafts that could connect to the underground passage, each time coming back empty-handed.

On July 1, Morris had hiked up with a military team to the south side of the mountain. They were hiking at the level of the river, which they knew drained out of the cave. They also knew they were already a mile south and at least 150 feet above where the boys were. A platoon of soldiers had been trying to widen a crease in the ground for days. When Morris wiggled in there he could tell immediately that it was a dead end. Cave entrances breathe, and this crease gave off not even a whisper of breath. The air was stale. This particular cave had likely collapsed 100 million years ago, he informed the officers. They insisted he keep trying. He told them that it might lead down, but anyone attempting the descent was likely "to climb through stacked boulders and end up in a *127 Hours* situation, Aron Ralston style"—a reference to the movie based on Aron Ralston's memoir of a bouldering accident that forced him to self-amputate his arm with a pocketknife.

The following day, Monday, July 2, Morris and his team had linked up with the U.S. Air Force Special Tactics team trying to clear the helicopter landing zone near a possible drilling site. They clicked. On Wednesday, July 4, as Stanton and Vollanthen were making their supply run to the boys, Morris was assigned to join a Thai general whose team had been scoping out a cavern for four days: Twenty men had hunted around a single cavern for four days burning hundreds of manhours. Once again he got there to find no air, no sign of depth, and no cool mist. Some of the crevasses they were checking out were only a foot wide, not to mention the fact that they were about fifteen hundred feet above the boys.

The assignment, though, was not a total bust. On that day he had come in contact with Thanet Natisri, the enterprising Illinois-based Thai expat who headed some of the rescue mission's water-management projects. They perused satellite images for dents in the topography which might lead to an alternative cave entrance, and planned to talk about their observations the next day.* On Thursday, July 5, the pair met at a coffee shop in the hotel in Pah Mee where many of the divers were staying. Thanet pulled a laptop from under his arm and showed Morris satellite images of promising sinkholes they could explore that day. Morris immediately liked Thanet. "Right away I knew the kid was switched on. He had maps and had done seismic studies."

They hitched a ride back to the mountain and Morris pointed at a cliff that corresponded to a possible sinkhole Thanet had shown him in satellite imagery.

"Can you get a chopper to take us there?" he asked.

Thanet happened to know a guy, as his benefactor at the camp was Colonel Singhanat Losuya from the Thirty-seventh Military District in nearby Chiang Rai. His unit had been one of the very first deployed in the search operation back on June 24 and he'd kept a finger on the pulse of the entire rescue operation.

Morris and Thanet met at base camp the next morn-

---

* Sinkholes are often formed on the surface when a cave ceiling collapses from gravity as acidic water eats away at the limestone above it. They are often excellent indicators of a possible cave shaft.

ing, Friday, July 6, for their requisite flight briefing at headquarters. For Morris the chopper ride was a bit of a lark. The chances of finding anything related to that sinkhole were almost nil, and since his skills weren't needed he'd been planning to leave that day at 2 P.M. for the four-hour drive back to Chiang Mai. He had rescheduled his flight to the United States for that coming Sunday, July 8.

As Morris and Thanet received their briefing, in the background Morris heard a voice he recognized. He turned around to see the commander of the Thirty-seventh and Colonel Losuya's boss, Major General Buncha Duriyapan—known as Big Black. Big Black knew Morris's wife; in fact, fifteen years earlier, before she and Morris were married, the now-general had taken the two of them out to dinner—mostly to vet the skinny foreign suitor. Morris was not surprised to see the general there; a few days earlier, before Morris left for the cave site, his wife was watching TV and had exclaimed, "That's him, that's our general!" So after the flight briefing at base camp, Morris stood up and reintroduced himself. The two briefly reminisced about that long-ago night, and the general introduced Morris to the rest of the leadership as his old friend.

In Thailand personal connections are critical and are known as "Sen," which means "noodle." There are three kinds of noodle that people talk about: "sen mee," which is like vermicelli—a tenuous connection; "sen lek," like fettuccini—a mid-range connection; and "sen yai," the half-inch-wide noodles one might find in the Thai dish *pad si yu*—connoting a robust connection with someone that is important. By introducing Morris to his fellow commanding officers as his old pal, Big Black had just elevated Morris's social status, giving him the "sen yai noodles"—i.e., pull—that would help him play a role in the cave-rescue drama that no one had anticipated.

On their way to the military truck that would take them to the helipad, Thanet and Morris ran into Vern Unsworth, John Vollanthen, and a British diplomatic official. The Brits wore a look of defeat. They told Morris and Thanet about the low oxygen levels in Chamber Nine. Vern said bluntly: "If we don't dive, everyone dies." But they said no one seemed to be listening to them. Thanet had already gotten an earful about the problem from his American friends and from Vern. But to Morris it was a revelation. As Vern and the Brits continued on to a planning meeting, Morris nudged Thanet, saying, "I don't think we should get on that chopper. I think we need to stay."

Indeed, Morris would end up staying for five more days.

First, Morris and Thanet turned around and went back to Colonel Losuya, the deputy commander of the Thirty-seventh. He was the military commander first on the scene on June 24, whom the parents had beseeched for help. He's the father of a boy who could easily have been a Moo Pa. He had been sympathetic to the poor drenched parents crying in the rain that first night, and he remained not only sympathetic but committed to their sons' rescue. When Morris briefed him about the conditions the boys faced in Chamber Nine, his jaw dropped. He had never heard any of this. The colonel's next few decisions would eventually derail his career, but he felt compelled to act.

"It's been more than three days since they found the kids, and nothing's happened," Morris told them. "We had no news from commanders. And information [was not being] shared. The Thai Navy SEALs were not sharing information with the UK divers. The UK team, the U.S. team, the international teams, and the Thais had no high-level meetings together."

Thanet had also been trying to tell people that, but for some reason on Friday, July 6, people actually started listening to this toothy American climber who spoke fluent Thai—somehow his foreignness and his "noodles" with Major General Buncha helped him circumnavigate the rigid Thai hierarchy. They huddled in a side room of the headquarters, and the colonel picked up his phone. He dialed a former military buddy who headed the prime minister's bodyguard. The security chief relayed the message to the prime minister. A few nerve-racking minutes later he called the colonel back: "The PM himself does not agree with you." The security chief told them that the government still considered drilling and piping in oxygen to the boys the best option. According to the three stunned people in the room, he said, "Stick to the original plan." The security chief told them if they had a problem to take it up with the SEALs.

While that answer sounded definitive, it remained unclear whether the Thai leadership had been informed that the drilling operation was a failure. A few dozen holes had been punched into the jungle, but none to any depth, and without seismic sensors it was like throwing darts at a map—being off by an inch was like being off by a mile. The distance separating the nearest drill bit and the cave was as great as the height of the Empire State Building. Maybe greater—no one knew for sure.

Similarly, Morris, Thanet, and Colonel Losuya couldn't be sure the decision makers knew that the operation to pipe in oxygen and wire a telephone line to the boys had effectively ended with Saman Gunan's death just hours earlier. The two reasons that had just been given for not taking action were both lost causes— the "original plan" had already failed, but no one in a position of power seemed to know that.

By then the flummoxed colonel had steeled himself to violate every cultural and hierarchical norm in Thailand. He left Morris and Thanet at camp headquarters and told them to wait there. Half an hour later he returned. He had lobbied his fellow colonels. They were willing to support him, but only if he secured the backing of the SEAL commanders. Standing outside the headquarters, the color washed from his face, the colonel grimly informed Thanet and Morris that he had no choice but to doorstop Rear Admiral Apakorn. "And you're coming with me," he ordered Thanet and a now visibly agitated Morris. Marching ramrod straight toward the SEALs' encampment, the colonel knew he was walking down a road from which there would be no turning back.

The colonel was about to smash military protocol— going to a commander three ranks above him, from an entirely different branch of the military, to push for what would be a political decision about something entirely beyond his authority. That's not how it works in Thailand's military (or in most militaries) and the colonel knew it was likely to end disastrously for him.

As they crossed the muddy encampment, brushing past the phalanx of unwitting reporters, Morris turned to Thanet and said, "Holy shit, I'm going to have to cancel my flight on Sunday." They were trailed by Thanet's two shadows, the pair of soldiers who followed him everywhere holding various maps and surveys. They passed through the checkpoint to the base camp of the rescue area—which by now had been cordoned off from the media by a one-hundred-yard-long green-mesh screen. They turned right and walked another forty feet or so to the SEAL tent. Inside, a pair of captains were busy with administrative work. It was as if a brigade commander of a National Guard unit based in Alaska had approached veteran U.S. Navy SEAL commandos in Virginia Beach. This was the SEALs' home turf, and they were running the show.

The colonel asked Morris, Thanet, and his men to stand back a few paces, and entered the tent. With his hands in front of him in a gesture of humility and supplication, the colonel asked the still-sitting captains for an informal chat with the commander. The military norms of lower-ranking officers deferring to higher-ranking officers was upended. The colonel knew these SEAL captains because his unit had been on the scene back on June 24, launching those failed initial search attempts as the first SEALs arrived. He was the one who had handed jurisdiction off to them, and they apparently trusted him enough to pass on the message to their boss.

The trio stood there not saying much for about fifteen minutes. Finally the commander arrived back at the tent. Morris and Thanet's eyes bugged when they saw the man. They didn't expect "the commander" to be the *rear admiral*—the guy on TV every day. The SEAL compound was part of a long series of event tents joined together in the shadow of the cave. You could pretty much walk from the Australian section, through the American section, and into the Thai SEALs' area without having to leave the shade of the tent. And due to the delicacy and scale of the breach of protocol the colonel was about to attempt, they had hoped for a meeting place that afforded some privacy, but the rear admiral showed them to a few plastic chairs in the corner of his operations area.

They sat down. Nearly knee to knee. The colonel began by thanking the admiral for the audience. The veteran SEAL sat stone-faced. He looked impatient, perhaps even a little bored. It had, after all, been a rough twelve hours. He had lost a SEAL. The nation was mourning and his oxygen hose project was a shambles. He had things to do. After introducing Thanet and Morris, the colonel asked Morris to explain the

British diving team's position: Oxygen in the cave was running out, therefore time was running out. They needed to dive the boys out immediately. Thanet added that an atmospheric river was threatening to pummel the area by early the following week. Waiting for the monsoons to wring themselves out meant waiting for the boys to die.

Apakorn was now paying attention. As this odd couple made up of an American climbing expert and Thai restaurateur/water-management expert spoke, Apakorn spotted a member of the King's Guard walking outside the fenced-off rescue area. He called him in. A few handpicked members of the King's Guard, composed of ex-military commanders, roamed the camp anonymously. Few there learned their names—they weren't supposed to know. During multiple interactions, the king's emissaries never introduced themselves to Thanet, Morris, or the colonel.

Called *Wongthewan*, which translates into "divine progeny," the King's Guard is an elite unit of army officers and former generals traditionally considered the leading military faction in Thailand—a country that prides itself on being the only Southeast Asian nation never to be colonized. Founded in 1870, it is Thailand's oldest unit and is tasked specifically with protecting the king and buffering him from other army units and politicians. The current Thai monarch, Maha Vajiralongkorn, had served in the guard himself during his tenure as crown prince. According to Paul Chambers, an expert on civil-military relations in Southeast Asia and a professor at Naresuan University, the king is known to be extremely close to King's Guard army officers. Says Chambers, "Since the present monarch was clearly interested in ensuring a rescue for the trapped boys, King's Guard senior officers were sent to act as the eyes and ears of the palace. Thus they alone could press for the 'go' call."

Morris and Thanet repeated to the King's Guard member what they had just told the admiral. As a top representative of the king—who was heavily vested in the success of the operation—he wielded enormous authority, and Colonel Losuya was visibly intimidated. He stepped away from the tent to make a call. Moments later he returned.

"I'm calling an emergency meeting. Now," he informed them.

The little group, which had tentatively walked the hundred yards or so from headquarters to the SEAL base, now crossed right back to where they had started forty minutes earlier—this time with an entourage. It was shortly after noon when they sat down in one of the war room's side chambers. The king's emissary dialed a number on his cell phone. Thailand's interior minister picked up. In the room were several representatives of the king, the lieutenant governor of Chiang Rai Province, the commander of Thailand's Third Army, a batch of colonels, Apakorn, Morris, and Thanet. There was some fidgeting in pockets as multiple people present searched for the record buttons on their phones. Some needed to document their very presence in the meeting, others needed to cover their asses in case things went spectacularly bad.

The bigwigs spoke first, then ceded the floor to Morris. To get close enough to the cell phone speaker to hear and be heard, several of those present had to play musical chairs, taking turns in the seat closest to the King's Guard officer. The interior minister, Anupong Paochinda, was all business. He had patiently listened to all the generals and politicians. Now he peppered Thanet and Morris with questions about the drilling operations and the possibility of running in air to the boys. Morris settled in.

"You have two terrible choices," Morris told him. "I don't envy your decision. In one, everyone is going to die. And in the other, some people are going to die." The minister and some of the politicians had pointed to the 2010 rescue of thirty-three miners trapped twenty-three hundred feet below the surface in a Chilean mine as a possible template. But Morris explained to the interior minister that it took sixty days to rescue the Chileans. Drillers knew precisely where they were located—plus, the miners had food, they had a safety room, and they had air. Those drilling teams had also worked on relatively flat terrain under clear skies. In sixty days, Morris continued, the boys will be dead, having either starved to death, been drowned by rising waters, or been asphyxiated by dipping oxygen and rising carbon dioxide.

"Okay, I understand," the interior minister said. He asked to speak to Rear Admiral Apakorn and ordered him to officially abort the already abandoned operation to route oxygen and a phone line to the chamber. Every effort was now to be directed at stocking the cave with air tanks for the rescue.

He told the group he was boarding a chopper and flying there that evening for a briefing on their rescue plan.

Rescue plan? Thanet and Morris didn't even know what that would look like.

Chapter 15
Getting the Green Light?

Thanet and Morris crossed the camp once again, heading down the muddy hill from headquarters to the base camp, through the fenced-off area, and into the American section of that giant operations tent. They told the Americans about their round of meetings with the Thai bigwigs, informing them that the interior minister was on his way and that they needed to form a detailed rescue plan.

"We've had a rescue plan for a couple of days," Derek Anderson told a noticeably relieved Morris and Thanet.

Indeed, over the previous thirty-six hours, Hodges, Anderson, the Brits, the Australians, and the Euro-divers had engaged in a blitz of planning; they'd come a long way from where they'd started Wednesday night, when they'd ruled out extracting all the boys in one day and accepted that the boys would all have to be sedated. At a subsequent meeting on Thursday, they'd changed tactics, beginning with their initial assumption of the impossibility of a dive mission. Standing in front of a blank whiteboard, "impossible" was a word that neither Anderson nor Hodges was willing to accept.

"Just to say it's not an option, we were not satisfied with that," Anderson said.

They gently prodded the exhausted British divers. "Okay, let's just say we can only get four to six [boys] out," said Anderson, acting as sort of a life coach trying to build a perfect world. "Better than not even trying. Let's try to whiteboard some plans. Talk about feasibility."

There was no template for this, but they started throwing out more ideas, with Stanton, Vollanthen, and Vern growing more enthusiastic as the scribbles began to creep across the whiteboard. First they drew a rudimentary map of the cave. Vern helped them calculate the distances and the possible water levels at different stages of the cave. Stick figures depicted the path of the divers and circles represented the air tanks built up in the different chambers. Two divers would take a single child—the lead diver would hold the child, the second diver would provide backup and maybe carry the boys' air tank. They crunched the numbers, assuming that a diver might need to change tanks up to eight times during the dive—in a casual dive for experienced divers at that depth, that much air could last sixteen hours, but this would be no casual dive. They figured they would have to stage the tanks in a series of key locations along the route—those "gas stations" that the Euro-divers had set up. But they would need more of them.

One of the hurdles was placing the tanks—not many divers in the world had the ability to stock tanks as deep as Chamber Eight, about twelve hundred yards in; another was how to know in the dark which tanks were full and which were empty. This distinction was a matter of life and death, because the diving community had become convinced that Saman Gunan had either been mistakenly given an empty tank or a tank accidentally laced with carbon monoxide. Anderson had an idea that roughly followed maritime protocol: spent tanks would be placed on the right of the cave and marked with red chem lights; full tanks would be placed on the left and tagged with green lights. Some of the chambers would look like they were adorned with a subterranean Christmas theme.

The planning had continued into July 6, when the two Australian doctors arrived. Their mandate would be to ensure that the boys were sedated throughout.

With the original estimate that only four to six of the boys would survive the journey out, survivability became critical. The Americans again hit the whiteboard, listing the ways the boys could die in the cave on the way out: drowning, if their masks came loose or became flooded; hypothermia from a weakened body's exposure to hours of cold water; even an inadvertent bashing of their heads against rocks. They decided the boys should wear thick wet suits with hoods to avoid hypothermia, and—most crucially—that they would wear specialized positive-pressure full face masks.

The positive-pressure full face masks would feed the boys air continuously—unlike regular masks, which provide air on demand when a diver takes a breath—to ensure the comatose boys kept breathing, and might also help purge flooded water from the mask. While the U.S. team hadn't brought cave-diving equipment, they did bring positive-pressure full face masks—four new ones and one slightly older spare. That meant the delicate masks would each have to be brought back to Chamber Nine at the beginning of each day's rescue.

But there was a hitch with those masks. If a mask's seal broke from being knocked around in those cramped tunnels, it might start to flood with water; the divers likely would not notice it as they navigated the darkened exit route. There would be no way to know whether they were towing a sedated kid who was dead to the world or a kid who was actually dead.

Since drowning was such a major concern, Anderson and the team had an idea. Divers typically breathe regular air stuffed under great pressure into a steel tank. With the exception of those undertaking deep or complicated dives, divers simply don't need added oxygen. Unless, of course, you're a kid shot full of a cocktail of drugs powerful enough that someone could perform open-heart surgery on you without your feeling a thing. Among the items the Thai SEALs had abandoned in the course of their doomed oxygen tube mission were a number of oxygen compressors, which can pump pure oxygen into a dive tank. The Americans would ask the SEALs to begin filling tanks with as much pure oxygen as they could fit. The rationale was that if the boys' systems could be saturated with oxygen, it might buy them more time in case a face mask flooded.

The world record for what's called static apnea— basically holding still in a pool for as long as possible on a single breath—is just under twelve minutes. The world record for static apnea in which a diver is allowed to breathe pure oxygen for up to half an hour before beginning a breath hold is twenty-four minutes. Twice as long—long enough to watch a full half-hour sitcom without commercials. Neither is a pleasant experience. After the first couple of minutes, as carbon dioxide builds up in your system, your body begins to alert you that it really wants you to breathe by issuing messages to your diaphragm to begin contractions. At first it can feel like a flutter in the belly, but the longer you hold your breath the more violent the flutters can be. After an hour's session with free-diving guru Kirk Krack in 2012, I held my breath for five minutes. He explained that the trick is "riding the bronco" of contractions— knowing that it's only pain—because your body can go for many minutes after those contractions begin.

An unconscious person is unable to ride the bronco and override the diaphragm's demands, and the body might naturally try for a gulp of air. If one of the boys suffered a flooded face mask it would have meant breathing water and the onset of drowning. Humans typically lose their dive reflex at about six months (the famous cover of the grunge band Nirvana's break-through album *Nevermind,* in which an infant is suspended a couple of feet deep in a pool, is an example of a baby displaying a dive reflex). The divers were hoping that somehow the boys' brain stems would trigger a mammalian dive reflex, just as it had in the subjects in those studies of pinnipeds.

Giving the boys almost pure oxygen would buy the

divers a precious few minutes in the event of a flooded face mask; they could then race the drowning boys to the next chamber, where they might be able to revive them—this assumes that the divers even detected anything was wrong. It was a plan, all right, but it wasn't pretty.

Then there was the problem of how they would carry the boys. Some of the bigger and more experienced divers, like American Bruce Konefe, had to quit the search because they were simply too big to fit through the restriction after the sump at Chamber Three. There was another tight spot farther on that forced a diver to go upright, in a move that was like sliding behind a curtain. Agile cave divers could navigate that. Doing it while carrying fragile boys would be different. And while swimming in open water with the boys would not be particularly physically demanding, hauling an inert boy over dips and rises strewn with eons of rockfall and boulders would be.

The planners decided to supply the Euro-divers with a flexible plastic stretcher called a Skedco, which wraps around a casualty like a taco. The Euro-divers would wait at prearranged locations in the cave—Chambers Seven, Six, and Five—load the boy into the Skedco, help the diver carry the litter to the next sump, and assist in refitting each boy with his mask and easing the diver into the water. Chamber Seven was about two hundred yards down from Chamber Eight; Chamber Six was to be located several hundred yards farther toward the entrance of the cave, after the diver would make his right-hand turn at the T-junction. Chamber Five, the final "gas station," would follow about 250 yards later. After Chamber Five the rescuers would be on their own for the final four or five hundred yards to the waiting medics and rope systems at Chamber Three.

"There were a lot of risks involved," said Mallinson of their plan. It was the divers' recommendation that "we ought to accept those risks."

They knew that resistance to such runaway risk at the outset would be significant, and that once the operation got under way there would be almost no tolerance for fatalities. But after all the ideas had been exhausted and the whiteboards cluttered, then wiped clean, then cluttered again, it was clear that this was the best shot they had to dive the boys out—and diving them out was the only way to save them.

But though they had spent days devising the plan, until Josh Morris and Thanet walked into the American tent on July 6, their primary audience had been one another. As Morris and Thanet listened to the Americans talk, they both understood the plan and the care with which it had been drawn up. They also understood the terrible risks involved.

Now they all just had to convince the *pu'yais* to green-light it.

If a bomb had gone off in that room, it would have taken out a sizable portion of the Thai leadership. They were in the "big" war room at ranger headquarters, which had become the command center. It was 9 P.M. on July 6. For many in the room it had been a long day. They were all there: the interior minister, the King's Guard officers, Governor Narongsak, his lieutenant governor, the generals, the SEAL rear admiral, the colonels, Morris and Thanet, the Americans. Anderson had brought in his ever-present whiteboard and surveys.

It was hot and stuffy. About forty people were packed in, stacked up to the rear of the room with those with less seniority leaning against the back wall. Officers sweated in their uniforms. The conference table, which was actually a congress of several white-topped folding tables, had water bottles and eyeglass cases on it, but no phones. Interior Minister Paochinda ordered complete secrecy. No recordings. No leaks. And yet moments earlier there had been another general fishing in pockets for phones with their various recording apps. This time it was less about ass-covering and more about posterity (a few photos and videos would eventually slip out).

Major Hodges was seated to the immediate left of the interior minister, who despite the heat was wearing a dark jacket and black shirt. An interpreter sat in a chair between them, completing an intimate little triangle. Hodges started by urging everyone to leave emotion out of the planning. It was a difficult thing for the Thais in the room to do because their careers and livelihoods depended on success. Hodges told the minister, "The environment in the cave is working against us, with degraded oxygen levels and the waters that are coming in. The flow coming out of the cave is strong but stable." He explained that they couldn't get enough food into the cave. "All of these things tell us that we have got to do something right now."

When it was his turn to speak, Anderson was even more blunt: "We either have a shot, where we could get some of them out, or we leave 'em in there. And there's a very, very high chance that none of them survive."

Diving sedated boys out through a mile of bone-chilling water had never before been contemplated, much less attempted; the optics caused significant discomfort among the leadership. Distilling the risk matrix his team had worked on, Hodges told the group there was a "very, very high level of risk and a very low probability of success." Governor Narongsak then calmly asked the American major to define "success."

The answer: "If we bring back just one boy to his parents, I'd consider that a success."

Narongsak, who had helmed nearly every press conference and who had repeatedly assured the press and the world that all the boys would be home safely at some point, was now presented with the possibility of having to inform the world that some of the boys had not survived. The admiral listened unflinchingly. "He now understood," Hodges later said.

Most in the room now grasped that an immediate rescue operation was necessary, but the brass needed to hear how it would be done. Anderson, who was sitting at the other end of the long conference room, began briefing them—going through the plan they'd crafted over the last two days. At one point the interior minister stood up, grabbed a chair, and dragged it across part of the room to sit an arm's length from Anderson. He waved off his translator and listened. When Anderson stood up to trace the route the rescuers would take, the minister also stood up; he peered closely at the hand-drawn sketches that had been meticulously annotated in English and Thai by Stanton's girlfriend, Amp. The gallery of generals and bureaucrats sat stone-faced; given the intensity of the minister's interest, it was clear the Americans and Brits were now the leadership's grand viziers.

They were at the zero hour, and yet even then there was pushback. After Anderson finished his presentation, a Thai official raised his hand and called for another analysis of the water conditions in the cave: "Why are we rushing into a rescue?" Others started raising hands, trying to press the interior minister; Major General Buncha, Josh Morris's ally, put a stop to it. The interior minister instructed the military personnel to afford the international squad whatever support they required. He gave them the green light to continue planning and preparing the cave. The official go-ahead would have to wait until the prime minister decided on whether to proceed with the actual operation. There was one more thing: this Friday-night meeting was considered top secret. The interior minister ordered everyone that "This information must stay in this room." He warned officials against leaking anything to the press. And nothing did leak.

By forging ahead with the plan for a rescue dive, the interior minister—the highest-ranking public official in the room—had just stuck his head into a guillotine.

A few dead boys could spell the end for him in a society where failure exacts a punishing toll. Almost every person who came out of the room that night considered his actions heroic.

As the group disbanded, members of the King's Guard and the interior minister asked Morris, Hodges, and Anderson to follow as they filed into a smaller side room. They asked them to do their brief again. A trio of Thai officials would transcribe it verbatim and relay it to the prime minister (and likely the king). "And we did the same exact brief again. And those scribes were writing down every word," said Anderson.

The minister told him: "Okay, that's—that's good. That's the best plan that we've heard so far. Go ahead and start planning as if you're going to effect this rescue."

It was decided that "stage one" would be the team's code word for all the organization, prepping, planning, and practice runs they could get in before Sunday. "Stage two" was the real thing. Once that was set in motion and the divers dipped under the canopy of rock at Chamber Three, it would be nearly impossible to recall them. The pu'yais now knew that, and it was one of the reasons the mission's traffic light was still at red and not at green.

Given the preparation needed, the team told the

pu'yais the rescue could begin on Sunday, July 8, at the earliest. It was Friday night, thirty-six hours before they hoped to send divers walking into the cave.

The interior minister took it up with the prime minister. The royal guards relayed it to the palace. And the world, including the boys' parents, knew nothing about it.

Chapter Sixteen
D-Day

Saturday, July 7, dawned bright. As it does nearly every day at that time of year, the heat had settled and towering cumulonimbus clouds packed with moisture reared up. They released it sparingly, the rain mercifully measuring only a fraction of an inch. Outside the cave, microbes and fungi on the forest floor munched on plant litter and animal waste. Insects labored away, carrying food to dens. Ferns, thorns, and flowers stretched their leaves to welcome the abundant sun. So-called whistling ducks yammered in the new lakes created by the pumped outflow of water from the cave. Wild boars rooted around, and the indigenous mountain goats called gorals scrambled up the steepest slopes.

At camp, it was not just any day. In a whirl of much-needed housekeeping, teams of workers began scurrying through the cave. Over the past two weeks, it had become a dump so littered with trip hazards that the Australian team's biggest concern had switched from drowning to tripping—specifically, the danger of head injuries from falls on sharp rocks. Next came the possibility of electrocution. All those pumps, lights, jackhammers, air compressors, and Wi-Fi repeaters required electricity. Initially, workers used only what they had—domestic power cords, often stuck down with tape. Since many of the cables were installed piecemeal, at separate times by separate teams, the result was a rat's nest of wires that ran from the mouth of the cave through several partially submerged tunnels and chambers to terminate about a thousand yards later in Chamber Three. Workers were routinely losing their footing on them.

And while Chamber Three could communicate with the outside world, had ample food, and was populated by teams of international rescuers, it bore one striking resemblance to Chamber Nine, where the boys were. It stank. Rescuers had been pissing against the wall for weeks—maybe doing other things as well. There were discarded food wrappers everywhere.

An Australian at the site joked that had this been

in the U.S., the entire operation would have been shut down by the Occupational Safety and Health Administration, soccer team be damned. But this was Thailand, and in the beginning local authorities threw whatever resources they had into the search; by now the full might of the SEALs, the regular army, and every politician was behind the rescue. So the workers tried to make sense of the wires, wrangle the miles of pump hoses, and generally organize the mess.

Richard Harris and Craig Challen slipped past all that housekeeping for their initial dip into the chilly waters at Chamber Three, en route to their first visit to Chamber Nine.

As it turned out, it would also be decision day for the boys. While Harris and Challen swam out to Chamber Nine, Dr. Bhak explained to the boys that the pair of Australians who were coming needed an answer to the question of whether they wanted to wait out the monsoons or dive out. They excitedly answered that they were ready to leave. No one wanted to stay in that tomb. It was a good thing, too, because the doctors arrived with a message they might not have been fully authorized to deliver: You are going out tomorrow.

In the dim light of Chamber Nine, the doctors examined the boys and quietly calculated how much sedative might be required. They found that some had symptoms of pneumonia, but they and their coach seemed otherwise relatively healthy, if rail thin. Key to the success of the operation would be the order in which the boys went out. The planners wanted the healthiest boys out first. A stretcher bearing a corpse would crush morale in camp. But the doctors felt that, given the boys' state of relative health, they and their coach should decide. They told the boys to talk it over that night and decide on the extraction order. Before leaving, they had something to give them. On the same waterproof paper their sons had scribbled notes with little doodles of hearts to their families, their parents had written back.

Night's parents wrote:

"Dear Night. Dad and mom are waiting to arrange your birthday party. Please get out soon, and stay healthy."

Adul's parents, Myanmar refugees camping out across from the cave with the rest of the parents, wrote: "Father and Mother want to see your face. Father and Mother pray for you and your friends, in order to see you soon. When you get out of the cave please thank all the rescuers. Trust in God." (His parents, like Adul, are Baptists.)

Titan's mother urged the little guy to be strong.

"I wait for you in front of the cave. You must make

it! I believe in you. You can make it. I'm giving you moral support all the time. Love you so much. Your dad also misses you and loves so much."

Many of the parents urged their sons to tell Coach Ek that they were not angry with him.

Almost every parent wrote a note. The single exception was Dom's mother. She had gone back to help her parents with their amulet shop. The Thai SEAL who was compiling the letters noticed the absence of a letter to Dom, and instead of calling the mother to ask her to return to camp or to dictate a letter, she says he jotted one down himself: "Dom, your mother is busy with work today." It had been a few days since Dom's mother had wept, but the tears dribbled down again: the first words her precious son would hear from his mother after two weeks of hell were that she was "busy"? Dom's grandmother, her elegant bangles clinking, held her daughter right there in the shop's office amidst the boxes of goods, promising her she'd be able to make it up to him soon.

All of the parents were devoted, but Dom's mother was perhaps more than most. She is broad-shouldered and sturdy. She says that once the Thai SEALs arrived she'd started lifting water bottles and anything else in camp that resembled a weight she could lay her hands on "to train." Maybe, just maybe, she thought, they would allow her to join the team going in to the boys. It's not that she was delusional; she and nearly everyone else outside the cave woefully underestimated the treacherousness of the mile-and-a-half route to Chamber Nine.

The boys and their coach, who had actually walked the now-submerged route, would have had little concept of how much more difficult it would be under water. They also had no notion of the "Cave Boys" hysteria that awaited them in the outside world, including the one thousand journalists just outside the mouth of the cave—and mercifully so.

That is why when Coach Ek was asked to decide on the extraction order, he thought about it for some time and came up with a plan. To make the arrangement as fair as possible, he told the boys that those who lived the farthest away would go home first. This logic stemmed from the presence of the bikes that the SEALs told them were still leaning against the rail near the cave entrance. He figured that—weak as they were—all the boys would have to ride their bikes home (he was the only one with no bike waiting for him). So Coach Ek instructed the boys that upon exiting the cave they should ride to the nearest food stall or market (which weren't far from the cave site) fill their bellies, and ride home. Dr. Bhak chuckled when he was informed of the plan. He had to admit that it definitely had a certain logic to it.

He had grown remarkably fond of the boys and their coach in their days together. Their vigor returned after he got a few meals into them. To pass the time and relieve the boredom, the boys and the SEALs devised games of chess and checkers using clods of dirt and rocks as game pieces. Everybody but Titan played. The youngest boy was afraid of losing—perhaps rightfully so. They were playing against Thailand's Alpha males, who were pathologically disinclined to lose—they weren't going to throw a game just because these boys were stuck in the bowels of the earth.

Still, the boys and their coach, as well as the Thai SEALs in the cave with them, were all shielded from the frightening truth: An elite athlete had already died in that rock-spiked warren of tunnels. If the boys made it out, they would be so heavily sedated that they would likely not wake up until they'd been ferried out of the cave, whisked away in an ambulance, and flown to a hospital. Blissfully unaware of all this, they continued to tuck into their dwindling supply of MREs and jabber about the proper Thai food they would eat when they walked out.

After Harris and Challen completed the grinding swim back out of the cave complex, Harris told Stanton that he was unsure he would be able to take a boy out on his own if it became necessary—he was still too unfamiliar with this particular cave and the myriad ways it could conspire to kill or entrap a diver.

As climber Mario Wild recalled, it was now clear that the American team had taken point in planning and organization; its message was clear: "We can do this."

A day earlier, on Friday, July 6, Wild had been reassigned from his duties vetting possible alternative cave entrances. The Americans wanted him and his team to look at something. They offered Vern as a guide. Wild had been toiling on the rescue for more than a week, but always fifteen hundred feet above the cave on the mountain. This was the Austrian caver and climber's first time inside Tham Luang. "It was super beautiful. Amazing. There was this big mouth and there is a river flowing through it." It took a fresh pair of eyes to see the wonders that this now-battered cave still offered.

They splashed through the tunnels between the first and second chambers. Days earlier the Chinese diving team had drilled in a high line—like a towering clothesline—to deliver tanks up and down the steep embankments from Chamber Two to Chamber Three. It had worked exceptionally well, taking out spent tanks and zipping in new ones. It spared the

Thai SEALs and soldiers from hiking up and down the sloppy mound separating the two chambers. To the right of that ramp was the drop of about forty-five feet that had so spooked Euro-diver Karadzic. But Anderson's main rope expert, Sergeant Sean Hopper, needed to know if the high line could support something bulkier and substantially heavier than air tanks: children. Specifically, children on stretchers weighed down with oxygen tanks and some other hefty equipment.

The Chiang Mai climbing team said, well, no. The bolts were not strong enough to support a rescue, not when human lives were at stake. So the next day—Saturday—Wild and Taw, who led the Chiang Mai rock climbing team into the cave on the first day of the rescue, were back inside. They used laser pens to determine the precise line of drop, careful to avoid the glass-sharp rocks that could easily sever climbing rope. With the help of the American team, they calculated the number of new bolts needed and the sturdiest parts of the rock in which to drill; within hours they'd bolted in a hardier system.

During their work with USAF Special Tactics rope experts they noticed that the bundle of pump hoses dipping down the ramp in Chamber Three looked so much like a Slip 'N Slide, already greased with mud, that maybe it could be used for just that purpose. The climbing team reckoned that sliding the boys up on a pulley system might be easier than carrying them up. They also helped create a rope system that would assist in fishing the boys out of the sump at Chamber Three.

The USAF team had created a sign-in sheet that required every person heading into the cave to check in and check out. There would be no more marooned workers and no more lookie-loos. If the initial days of the rescue were defined by heart and chaos, the planned rescue itself had to be defined by order and organization.

The team was now fully assembled. Beyond Chamber Three there would be a total of only a baker's dozen divers, upon whom the rescue would hinge: the two Australian doctors, Challen and Harris; the four lead British divers, plus three support divers they had enlisted—Jim Warny, Connor Roe, and Josh Bratchley; and the four Euro-divers (down one after Vollmar wrecked his equipment on that first dive and dropped out). The number of rescuers equaled the number of people to be rescued.

But perhaps the biggest challenge would be communication—there would be none of it beyond Chamber Three. The only plausible way of dispatching messages would have been sending a runner (swimmer) from chamber to chamber—but that would risk head-on collisions underwater. So each actor in this saga had to memorize his role, timing, and exact placement along the mile-long stage that was the route out.

Anderson had a fix for that: "I was talking to the Thai SEALs. 'You know,' I said, 'all this has been great on paper, but if we have the ability, is there a pool nearby here we can use? We have the kid-sized wet suits. We have the full face masks here. Like, let's go try to run this to the ground and at least see how the equipment works. Let's see if the full face masks seal. Let's see if the wet suits fit.'" And with that, the SEALs scrambled to recruit a few local boys for a wet run at a local school pool that they kept top secret.

That Saturday morning, with Harris and Challen doing their first test swim in the cave, the group in the practice pool was intentionally kept small: a few Thai SEALs, members of the U.S. team, the British divers, and a few of the Euro-divers. In the background of the videos shot that day you can hear the recreational swimmers at the pool chattering amiably among themselves. The rescuers still had to preserve secrecy; they didn't want the media finding out—plus, they were still not even sure the mission had been approved.

Some of the boys were on the Wild Boars team. The SEALs suited them up in the wet suits that would soon be worn by their trapped teammates, then eased them into the pool to practice fitting the boys with full face masks so that the seal would remain watertight even if jiggled slightly. They strapped oxygen tanks to the boys and had the swimmers cruise around the pool with them, trying different placements of the tanks. Part of what they had to learn how to do was to "sink the boys" so they wouldn't start bobbing to the surface like underinflated pool floats. But an element of this exercise was also ass-covering, so the teams could tell their superiors that at the very least they had tried to the best of their abilities to do the closest possible approximation of live practice.

Later that afternoon, as SEAL teams continued stockpiling tanks in the cave—including the special tanks filled with 80 percent oxygen designated solely for the boys and their coach—outside the cave there was movement. At a flattish area near the main headquarters, people climbed into cars, started them, and began a Tetris-like maneuver to repark them. They cleared an area about the size of a full basketball court. Huge green tarps were slung up like curtains across the hill to prevent reporters from peeking in—it was the start of a period of increased secrecy, which to us reporters on the scene was a clear tell that something was up.

Inside the cordon, Anderson's team and the SEALs

carried up red plastic chairs and more than ten cases of water bottles. They began building a mock-up among the mud and tire tracks—a scale plan of what the rescue would look like. They used rope tied to chairs to map the route and each specific chamber. The distances between each chamber and the number of air tanks needed at each place were marked by letter-size paper in plastic sheets taped to the rope. Several hundred water bottles were placed beside chairs denoting chambers. Bottles with blue tape represented air tanks, bottles with green tape meant tanks with oxygen earmarked for the boys, and bottles with red tape stood for empty tanks. In military parlance this is called an ROC—or Rehearsal of Concept—drill.

This was a military operation whose primary operators would be civilians. The two worlds don't often play nicely. It would be the first time all of the separate teams would work together. For several divers, English was not their first language. They had previously been working as independent cells, each dealing separately with the U.S. Special Tactics team, but now they were compelled to work together, sharing the same space. And at first, some of the divers considered the exercise ridiculous.

"I thought it was only something the military did in the movies," said Karadzic. "I couldn't believe they actually do this. But [Major Hodges] told me it's standard procedure."

For the next couple of hours the eleven foreign divers walked through the primitive diorama; after the first round, the titters faded. During his turn, each diver went through picking up water bottles placed on the left where they thought they would need a swapout and placing a "spent bottle" to the right. They also practiced their timing—trying to avoid nasty collisions of groups of divers in confined spaces. Hodges and Anderson hoped it would quickly build the divers' muscle memory. One of the most confusing pieces of choreography was sorting the two types of cylinders—air and oxygen—while keeping an accurate inventory of the number of full and empty tanks (and ensuring they were separated and marked) at each "gas station."

In their first walk-through they discovered a potentially disastrous flaw in the plan. Their designation of "chambers" was fluid. The only "chamber" designated on Vern Unsworth's and Martin Ellis's surveys of the cave was that hangar-size first chamber. The others, particularly those after Chamber Three, were basically arbitrary designations the SEALs and international teams had assigned to slightly larger spaces or openings along the cave's main tunnel after the search-and-rescue operation began. The idea was to standardize everyone's understanding of specific locations in the cave. For instance, all the divers knew how to pinpoint Chamber Five on a map. But put them in a walk-through of a simulated cave route, and that changed. This was partly because there were no landmarks demarcating those "chambers" in the cave itself; for instance, Chambers Eight and Seven were arbitrary terms for a stretch of mostly dry cave somewhere between Chamber Nine, where the boys were, and the T-Junction. The support divers who were to be staged at Chambers Five, Six, and Eight had different notions of where those rooms existed. They could be off by the equivalent of several city blocks—a continental divide in those darkened tunnels. If that happened during a rescue, it could lead to a fatality. So the team decided that Karadzic and the Euro-divers would move their "Chamber Five" one hundred yards closer to the T-junction than the spot they had originally scoped out. They also weren't sure if the instructions to stage in a certain chamber meant, for instance, that Rasmussen and Passi should stage at the sumps between Chambers Seven and Eight, indicating the flooded areas that connect them or in the heart of the dry areas proper. The people who drew up the plans had never been that far into the cave—for that matter neither had the Euro-divers—which meant that even if they were right for the purposes of the drill, they had no idea whether that correlated to the actual reality in the cave. The drill was repeated until everyone felt more or less comfortable.

The drill had a secondary purpose. Paasi, another ROC skeptic at first, soon realized that they were making a big physical statement, right on the doorstep of the headquarters brimming with generals and politicians. "We were showing everyone what we are planning, that it's happening, and this is the way we are going to do it."

It succeeded in its intended effect. A crowd of curious officers and politicians gathered around. Within half an hour there were dozens of spectators, including the interior minister, who walked over to Hodges and Anderson and stood beside them. They asked him, "Do we have a green light?" He immediately answered: "Yes, absolutely. Move forward with this plan."

Some had emerged from the previous night's meeting thinking the go-ahead had been granted; the operations planners, Hodges and Anderson, were not certain that was the case, and the interior minister's blessing was the final authorization they'd been waiting for. And so, about twenty-two hours before they needed to launch the mission in order to have a chance of beating the rains, they had official permission from the highest levels.

The media knew nothing and had little actual understanding of the remarkable efforts going on behind that green curtain separating our encampment and the screened-off operations area. So far there had been no leaks—in fact, even U.S. government officials working closely with the U.S. team were kept in the dark.

That Saturday, July 7, the ABC News Pentagon Bureau broke a story about an internal U.S. military report listing many of the rescue mission's details and intimating that it was close to happening. (This may have been another quiet measure to prod the Thai government into deciding on a rescue mission.) That afternoon I got a tip from a confidential source that the dive operation was about to start. I was told the boys would be taken out by buddy teams, two divers to a boy. The operation would take three days. They would be wearing full face masks. The water levels had declined to the point that most of the way would be above water. The majority of this information dovetailed with the Pentagon report. But in trying to confirm the information that evening, I spoke to a trusted American official just hours before the mission was to begin. He waved me off—telling me a dive rescue would be "extremely unlikely," a phrase he used multiple times. He called the diving rescue concept "extremely risky" (which was true), because rains could trap rescuers, creating an even bigger disaster (also true). He thought boring a relief shaft to the boys was still the option the Thai leadership would most likely choose and that the U.S. team was working in support of that. This was less than twelve hours before the mission was set to begin.*

Further confusion came that Saturday afternoon. Walking up to the main journalist campsite outside the mouth of the cave, I ran into the Thai minister of tourism and sport, Weerasak Kowsurat, who had been involved with the early decision to call up the British divers but was not present at that Friday night meeting. We were standing on the mud-slicked road, just down from one of the many pumps shooting water from the cave into a canal. He told me he had just come back from testing an inflatable tube concept that might preclude the boys from having to "swim" out. He said he had tested it out himself, crawling into and out of a one-hundred-foot section in a nearby pool to test its durability. He listed a number of challenges, including keeping what would effectively be a mile-long sock-shaped bounce house inflated for long enough. "We had two blowers that provided air for one hundred feet, but that only proved that it would work over that

---

* I am confident the source was speaking in good faith and was not attempting an elaborate misdirection.

distance." I asked him somewhat incredulously if this meant a diving operation was precluded. "I have no idea. I do not have enough information about the conditions inside," he said. He added that he had no idea how much time rescuers had to work with.

The impending mission was, in fact, kept so secret that many U.S. officials working closely with the American team had no idea it was about to commence. This explained the cases of reportorial whiplash journalists like me had just suffered. I'd received multiple reports from trusted sources that were so utterly contradictory that we elected to report none of them.

And as we waited, so did the parents. They had been informed that afternoon that a rescue dive would begin the next day, and were instructed to keep it secret. They were furnished with no details about the plan itself—not a single one of them knew the boys would be sedated. It wouldn't have mattered anyway. The Thai government was not asking for their permission, it was informing them what would happen with their sons. It was the only way it could have worked. There was no time for debate or dissent.

There was still work to do that Saturday night. The Chiang Mai rock-climbing team had to take a second look at its calculations and ropes. They hadn't managed to finish everything and needed to go back in the morning to tension some ropes and drill in bolts for hand lines that rescuers would be using to steady themselves along the uneven terrain.

But eventually everyone called it a night—Sunday was the big day. Rope rigger Mario Wild was nervous. Diver Jason Mallinson lay in bed mentally rehearsing the next day's movements. Thanet wondered if all his toil trying to wrestle northern Thailand's bountiful water into a more forgiving state for diving would pay off. Vern—whose knowledge of the cave had been seminal in helping map out the rescue plan—and his partner, Tik, lay in bed in their bungalow, with the Sleeping Princess a dark shadow beyond the dimly lit valley. The same questions rattled around his brain over and over: "What is happening with the water levels? Will it be pissing rain outside in the morning? Will it rain long? Can they do it? Will the divers even be able to get in?"

Anderson trudged back to his hotel on Saturday night, exhausted. He'd spent the evening trying to manage expectations. He still had no idea if this concept would work. They had practiced and they had prepared the best they could; now it was out of his hands. He had about five hours before he had to be up again, so he scarfed down some pasta and racked out.

Vern was up shortly after 5 A.M. He had only slept a little over three hours. It had rained early that morning—not a monsoon punch, but a steady rain accompanied by a distant growl of thunderclaps that lacquered Mae Sai and its surrounding jungle in a shiny shade of emerald. Pretty, but not the start the divers had hoped for. Vern and Tik drove from their home to the 7-Eleven to pick up coffee for Stanton, Vollanthen, and Harper. As they sipped the convenience-store lattes and ate some packaged food at Stanton and Vollanthen's hotel in Mae Sai, they talked logistics and the rain, which had started to slacken.

Reviewing it once more with Vern, Harper, and Vollanthen, Stanton felt confident about their logistical plan and comfortable with the group they would be working with. The forecast for the rest of the day looked decent: no major rain was predicted aside from the expected sprinkle. A veteran of six major international rescue-and-recovery missions, Stanton mentally girded himself for the day ahead and the likelihood that at some point his water-pruned hands would be carrying a dead child. It would hardly be the first corpse he'd handled, and he'd necessarily grown philosophical about the unpleasantness inherent in rescues; he explains in his characteristically blunt way that he cherishes life and would risk his own to preserve another's, but once a human expires, "I just treat it as a lump of meat. The human ceases to exist. It's just a shell the human once occupied."

Human corpses had never really bothered him. Even before he had joined the UK Fire Service in 1990, he'd come face-to-face with one of those shells "the human once occupied." It was May 1986, and he was a twenty-five-year-old who had been called to Yorkshire, where cavers had become stranded in a near-vertical cave named Rowten Pot, a giant hole in the ground atop a nearly bald hill covered in tweedy grass. Rowten Pot has a deep section that requires a complicated abseiling descent past a jet blast of water from a stream careening off rocks as it crashes down the hole. Two cavers had gone missing, and the area's Cave Rescue Organization was called in to save them. One of the rescuers was David Anderson, who had apparently slipped during his descent and fallen into a gully. The water came up so fast that he was trapped on his rope and drowned. And there he remained for the next several hours until Stanton and members of another rescue team rappelled down to recover his body. By the time they arrived the flash flood had subsided, and Anderson was found slumped in his harness in full kit. For the next several hours the team had to manhandle the inert body up several hundred feet of rope. They had come into unusually close physical contact with a dead person—one Stanton had known personally. It was the first corpse he had ever seen, yet he was mostly unmoved.

At the Tham Luang cave that morning, head park ranger Damrong, the man who had initially found the boys' bags more than two weeks earlier and had since overseen the registry of over one thousand journalists, would now have a hand in banishing them. Clearing the mud pit members of the media called home was the mission's first action. A sparsely attended "press conference" featuring a lone police officer had started, but it was poorly attended and conducted entirely in Thai. The intent became patently obvious when soldiers started helping journalists fold up tables and take down tents. ABC's internal WhatsApp page began crackling with warnings: "Dude, all the local media are packing up. Tents and everything. It sure looks like this place is being evacuated."

Everyone was ordered to be out by 9 A.M. By 8:30 A.M. the main paved road leading to the rescue site, the only road in, was blocked by military checkpoints. Minutes later a graphic with a map popped up on many journalists' phones, asking them to congregate at a donation center on Route 1 over a mile from the cave site. At the hotel thirty minutes from the cave where we were staying, a call to quarters sounded. Producers, reporters, and translators started scrambling to decipher what this all meant. Our coordinating producer, Brandon Baur, caught the eye of our team's lead driver, Nop—a big lug of a man with a long, sloping forehead and a head shaved bare. Nop was watching something intently on his phone. Baur tried to ask him what was going on. Nop started waving his hands and saying, "big news, big news!" ABC correspondent James Longman grabbed his phone, opened Google Translate, and put the phone to the driver's mouth. He rambled on excitedly in Thai for a full minute. He finally stopped, and Longman pressed Translate. Suspenseful silence. And then the phone spoke: "Never again, alpaca." They all started cackling. The tension was broken.

Something was afoot, and it wasn't an alpaca. The thousand members of the media mustered from their guesthouses and hotels. By 10 A.M. I was at the local government offices, which had been turned into a donation site, waiting for a press conference. A dais had been set up with four chairs under the eaves of one of the buildings. Hundreds of reporters clustered around a collection of blue tents with semicircular awnings, scrapping for chairs and space. Dozens of cameras targeted the little dais. There was no sign of Narongsak and the other speakers—it would be a while, a local official warned me. With time to kill, I walked around the little square. The kitchen crews had set up camp, doling out rice bowls crammed with vegetables and savory pork, energy drinks, and water. I walked through the government building, where industrial refrigerators held sacks full of pork parts. Out back, a platoon of elderly women hacked away at bok choy, onions, and garlic.

Over an hour later, this new journalist camp jolted to attention. Governor Narongsak arrived, flanked by Rear Admiral Apakorn and two other commanders. He sat down and talked—a lot. Anticipating something enormous, possibly the seminal moment of the rescue, ABC's executive producers back in New York had scrambled to prepare for a live special report. I was in front of the camera; crouching beside me was one of our translators, struggling to keep up with Narongsak. My job was to distill his translation and repeat it into the microphone for the folks in New York. The suspense was torturous.

Narongsak's preamble began: "We have been preparing for the main mission in every single way possible, and today is the same. It is the fifteenth day of the operation. . . . The search was like finding a needle in a haystack, but the rescue is proving much harder because the conditions we face are not normal. Nowhere in the whole world have people ever faced conditions like this before." He then discussed the mighty international commitment, the foreign divers, the NGOs, the regular volunteers.

"People outside would never understand the difficulty," he said, launching into an explanation of the continuing work on top of the mountain. "First we found the kids and now we need to get them out to the cave entrance. We are still discovering cavities; we are always drilling, and I can officially say that we have drilled more than one hundred cavities and dug into eighteen cavities with potential [but with no success]." There was discussion of the diminishing oxygen levels and the rising levels of carbon dioxide. Time was running out.

"We are still in a state of war against the water. All the plans must not have any holes in them, but there will always be margins of error." Waiting until December or January, after the monsoon season, he said, was now impossible. The next best thing would be reduced water levels, allowing an operation to commence. "So today there are only two plans, but many methods. We'll choose the best method to move forward with." At this point the reporters there were completely mystified. Producers in New York started asking in my ear: "So no rescue? What's he saying? They haven't decided which plan to go with yet?"

And finally it came out: "Today is D-Day." The rescue mission had begun at 10 A.M.—two hours earlier.

Wait, what? It started already? Two hours ago!?

Chapter Seventeen
Like an Egg in a Rock

The secrecy on the part of Thai and American officials had been complete, even if the governor's
description of the mission had not. Narongsak and
the generals told reporters that a buddy team of two
international divers would extract each boy, and that
this would take two to four days—that part was in fact
correct. Regardless, the divers, the SEALs, the Americans, the rope specialists, the water-management specialists, the Australians, and the Chinese gathered at
the mouth of the cave early that Sunday morning enjoyed the quiet. The journalists and some of the other
auxiliary volunteers had been flushed out. It was now
just them, the boys, and the mission.

A few hours earlier, before the press conference that
had initially confounded the media, the day's mission

had gotten under way. Up at headquarters—the old park ranger cabin—Hodges and Anderson had led a team briefing. The first part had included the whole group: the governors, the interior minister, the generals, the Thai Navy SEALs, the Chinese rescuers, the climbing teams, the U.S. pararescuers—everybody. For about forty-five minutes they walked through the concept again and asked if anyone had questions. Then all were dismissed to begin gathering the tons of gear that would be needed that day. The twelve divers, Anderson, and a few others split off for a smaller meeting. Dr. Harris had them sit around the conference table where thirty-six hours earlier Hodges and Anderson had managed to convince the Thai interior minister to initialize the mission.

Dr. Harris had something for them; each of the divers got a small pack filled with syringes and needles. He had spent the previous night tinkering with the right mixture of drugs for the sedation, jotting down notes and conferring with Thai doctors. He also had had a long conference call with Australian anesthesiologists and pediatric psychiatrists. Together they had agreed on a cocktail of drugs, beginning with Xanax, which would be administered about half an hour before the boys walked down to the water's edge; once there they would get two injections, ketamine and atropine.

The former is a fast-acting sedative frequently used in pediatrics, and the latter would be administered to dry up the boys' mouths and lungs so they wouldn't choke on their saliva or mucus.

At some point overnight between Saturday and early Sunday he had presented this final plan to Thai authorities and received the official green light to administer the sedatives. Without permission to sedate the boys, everyone had agreed that the mission would have to be aborted.

But the consultations overnight had complicated the plan. It turned out that it couldn't just be Harris administering the sedatives, because ketamine sedation wears off after about twenty-five to thirty minutes; the divers, particularly the four who were diving out the boys themselves, would have to readminister the ketamine— probably more than once during the rescue—to keep them unconscious through the entirety of their journey out. In other words, the divers would have to become amateur anesthesiologists, pronto.

Despite the risks of deputizing the divers as anesthesiologists, this combination of drugs seemed the best option; indeed, a great deal of thought and strategy went into the drug selection. To smooth their way into unconsciousness, and to limit the possibility of cold feet or panic as they approached the water (and syringes), the Australian psychiatrists had suggested that Harris first give them the benzodiazepine Xanax.* The drug stimulates the GABA receptors in your brain—the brain's natural antianxiety medication. By increasing the activity of GABAs, the half-milligram dose would rein in a mind whirling with fear.

Ketamine, as Harris explained to his team, was a little trickier to administer. By guesstimating each boy's weight, Harris would have to make a judgment call on how much to inject. There is a vast difference between a 70-pound eleven-year-old and a 130-pound sixteen-year-old. Injected intramuscularly, the drug would knock the boys out within one to five minutes, shutting down the region in the brain that notifies the body of pain. It has the additional side effect of acting as a muscle relaxer, which would limit involuntary movements of the arms and legs. Unlike its pharmacological cousin propofol—which killed Michael Jackson—ketamine would not arrest the boys' respiratory systems. In fact, ketamine has so little effect on the respiratory system that it's actually used in emergency rooms to treat se-

---

* It was actually the generic form of Xanax, called alprazolam, but for the sake of simplicity, let's just call it Xanax.

vere asthma cases and to combat panic, turning off the spigot of emotions like fear, pleasure, and anger that are controlled by the limbic system.

Those same properties, however, can make ketamine deadly when a patient is not constantly monitored, as would unfortunately be the case with the boys. It relaxes all the body's muscles, including the muscles that help us breathe and those that cause us to reflexively swallow. In so doing it allows secretions to build up in the lungs. Those secretions in the lungs, plus unswallowed saliva, can be a choking hazard.

That's where the atropine would come in. It would dry up those secretions—an absolute necessity, for the multitasking divers would be too busy following the guideline and avoiding rocks that could brain their young wards to check on the amount of saliva and mucus building up in their mouths.

On paper, the drug regimen made sense, but there was one sizable hurdle: the journey back from Chamber Nine would take roughly three hours, which meant that all the boys would require at the very least one additional shot, and quite possibly up to six more over the course of the extraction. If the boys woke up, Harris told them, the amateur anesthesiologists would have to knock them out again. It was crucial that they do so. An alert boy, coming out of sedation into the cold blackness with a mask tensioned on his face, would likely panic and could possibly suffer severe trauma and PTSD.

Each diver was given a pouch with several syringes loaded with ketamine. This drug is the most commonly used animal tranquilizer, which made veterinarian Craig Challen already accustomed to administering it, but all the others would be first-time anesthesiologists. And they were getting this news mere minutes before the rescue operation was due to commence.

Harris gave the divers a crash course in ketamine. He sat at a table, upon which were arrayed zippered nylon packs the size of small toiletry bags filled with syringes, needles, and several water bottles. He explained to the nervous divers that their mission was going to become slightly more complicated. While all of them had basic first aid skills, aside from the doctors only Rasmussen had practiced any kind of field medicine, when he had worked for the Red Cross years before. In a briefing that lasted about as long as it takes to read a magazine article, the divers were taught the basics of administering intramuscular shots. They weren't going to practice on themselves—thus the water bottles. Dr. Harris did it first, instructing them that needle control and positioning are key. You have to get to the meaty part of the body—the thigh is best and most likely to have enough muscle left on it after two weeks of short rations. You don't want to hammer it in, but you do need to achieve depth to ensure it plunges through the wet suit. Get it in the flesh, and punch down—don't worry about gas bubbles, because this is not going into a vein—and anyway, muscle tissue will absorb any air. Each diver got to "sedate" a water bottle or two. The four main rescue divers had received a similar briefing late the night before. But that was it.

The divers were clearly skittish. During the short training session, Harris assured them that they wouldn't be able to overdose the children, though that was not necessarily true. According to the FDA, "Respiratory depression may occur with overdosage or too rapid a rate of administration of [ketamine], in which case supportive ventilation should be employed. Mechanical support of respiration is preferred to administration of analeptics"—but supportive ventilation, and certainly mechanical support of respiration, would not be options during the upcoming dives. It was a potent cocktail of drugs, and Harris knew that the risk of administering it to children in a severely weakened state was enormous, but there was no other choice.

"We were a little surprised that this was how Harry was telling us to do it," said Rasmussen, who—like all of the Euro-divers—works in Thailand. They had not negotiated for immunity—hadn't even thought of it; the Euro-divers, arguably more than anyone, would face life-altering consequences if a boy died or was overdosed. They feared they could be arrested, lose their visas, be separated from their families, or lose their businesses. There was no time to consult lawyers or families; anyway, details of the mission were strictly top secret. Choking down his fear, Rasmussen told the team, "Okay, if we have to do it we will do it. The downside would be a kid waking up [during the rescue]."

Despite this last-minute wrench, the divers were increasingly confident they could execute their plan—the wild card remained the boys' reactions. As they headed toward the mouth of the cave, Rasmussen revised his earlier calculation about the boys' survival rate: instead of 80 percent fatalities—that is, only one in five boys would live—he now thought that one in four of the boys might survive the journey out. For the divers, rescue protocol and personal survival itself could necessitate an agonizing decision, which veteran diver and first-time rescuer Chris Jewell thought about this way: "I wasn't 100 percent confident that I would come out with a live child. I was confident that I could get out of that cave myself, but I knew there could be a scenario that would force me to let go of the child and sort of have to abandon him in order to save myself. And one of my biggest concerns was that in order to get myself out I might inadvertently damage or kind of displace the full face mask that was ultimately keeping the boy alive." That particular worry burrowed itself into the rookie rescuer's brain and refused to leave.

As the rest of the teams waited, the thirteen divers turned up the stairs for the hour-plus slog to Chamber Three. After them came a funereally solemn procession of over one hundred people.

"We were nervous," said Major Hodges. "My biggest concern was that we didn't have any communication capability past the third chamber. So once that part of the mission started, it was, like, we're hitting the Play button, and we're just sitting back and thinking hopefully we've planned this well enough. Hopefully the contingency plans are solid enough. Hopefully everybody back there knows what the plan is, the way that we had rehearsed it. But yeah, we were nervous. But then it was just a long waiting game."

Rear Admiral Apakorn told his superiors that the boys would be cared for "like an egg in a rock," which is a Thai idiom referring to a chicken that lays an egg and wonders how much safer its precious progeny would be if it were encased in stone, rather than in a fragile shell.

The mission had officially started, yet the teams still hadn't fully finished their preparations. The USAF Special Tactics team was still lugging equipment in for the Euro-divers and others and checking and rechecking all the rope systems they would be using in Chambers Two and Three. The Euro-divers still had tanks to place. And after the previous day's misunderstanding over the location of Chamber Five, they would have to find the cache of tanks and equipment already there and place it one hundred yards closer to the T-junction. Along the route to Chamber Three, Thai Navy SEALs and others peeled off to take their positions. The deeper into the cave it snaked, the smaller the procession became. The flayed feet, the hands so infected that Vollanthen and Stanton had to routinely lance their wounds to let the pus ooze out, the debilitating cases of jock itch, the mysterious rashes welling up beneath wet suits, and the ailments divers were calling Tham Luang foot rot and Tham Luang lung—a hacking cough that lasted for weeks—were all ignored.

About ninety minutes later, the Americans, Thai SEALs, and others were assembled in Chamber Three. Stanton was making a last cursory check of the equipment and the tanks they would take in, running his hands over them, when he brushed something that felt wrong. It was the oxygen tanks for the boys. At the top of any dive tank is a cylinder valve, which controls the amount of gas coming out like a spigot—be it oxygen or other mixtures of gas. Those valves have two different types of connection systems: an A-clamp connector and a DIN connector. Completely different sets of regulators and gear go with each—the difference is as big as that between a left-hand-drive car and a right-hand-drive car. The SEALs and the international teams had decided early on to standardize all their equipment. They would only use A-clamps for every one of the hundreds of air tanks prepared for them each day. But today, the first day of the mission, Stanton noticed that the oxygen tanks that would be used by the boys were fitted with DIN connectors—the wrong kind. Had they hauled them out to Chamber Nine, the divers would have been unable to connect them to regulators—meaning the boys would have no oxygen and the tanks would have been as useful as thirty-five-pound paperweights. The mistake set off a frantic search for the right connectors. There was no time to rush out of the cave, so Stanton and Mallinson decided they would cannibalize the right connectors from their own supply and fit them on the boys' oxygen tanks. The procedure ate up forty minutes.

Without fanfare, the American and Thai teams in Chamber Three watched the thirteen main divers slip into the water, each spaced a few minutes apart. They swam in file, but they swam alone. Cave-diving rescues are the most solitary kinds of rescue missions, and over the previous week most of the divers had spent more time inside the cave, alone, than they had outside it. For many, the shock of the cold water had become as familiar as their chronic fatigue. They weren't bothered by the murk anymore—work in which they called diving blind or diving by Braille. Many had committed parts of the route to memory. Stanton or Mallinson could often anticipate a slack guideline leading into dead-end wrinkles in the tunnel, and they knew when to deflate their buoyancy vests as they glided toward some of the lower areas. Challen, Harris, Stanton, Vollathen, Mallinson, and Jewell went first. The Euro-divers and the BCRC support team of Roe, Bratchley, and Warny followed.

Loaded up with air tanks and extra sedatives they prayed they wouldn't have to use, each solitary diver began navigating one of the most challenging parts of the cave, which began right after the sump at Chamber Three and included the straw-tight, 150-yard-long passageway leading to Chamber Four. "It's an awkward bit of underwater passage. There are several particularly tricky line traps, where the guideline we're following can get pulled up into small cracks and make it difficult to navigate," said Jewell. For most of that section the divers would have to slither— their bellies on the cave floor, their backs bumping its cheese-grater ceiling. It was just big enough that they wouldn't have to take their tanks off. And it was that dreaded part of the cave that would soon lead Jewell into trouble.

The Chiang Mai climbing team was among the groups holding the fort back in Chambers Two and Three. It was mission day, but they still needed to tighten a bunch of bolts and ensure that the lines were free of snags and the hand lines were firmly in place. There were to be no wildcatters today. Every single person in the cave was assigned a specified station and a specific task, with orders not to leave their posts. The entrance to the cave had a sign-in sheet and a board enumerating the exact number of rescuers inside at all times.

In that day's briefing they had also been given some auxiliary instructions. Said Wild: "We knew from the first briefings that some kids would not make it. We were to be prepared inside, if someone arrived dead, we would keep that person in there. Because they didn't want the optics of bringing out a dead body."

Then everybody waited.

About seven hundred yards from Mario Wild and his Chiang Mai rope specialists, Euro-divers Ivan Karadzic and Erik Brown had spent the previous half hour relocating what they had thought was Chamber Five to a new location one hundred yards away and double-checked that all the tanks were in order. It had taken them about an hour and a half to get into place and thirty more minutes to put the finishing touches on their work. And now, for the first time since Major Hodges had effectively commandeered them to his team four days earlier, they had nothing to do. It was "a bit of a weird sensation sitting there," recalled Karadzic. To conserve battery power and to better see the lights of incoming divers with kids in tow, they had dimmed their own lights so low that they could barely make out each other's faces.

"It was just me and Erik," said Karadzic, "and it was the first time in a long, long time that it was quiet, no stress. And we looked at each other and wondered what the hell have we done. We are two kilometers [about a mile and a quarter] inside a cave. We were gobsmacked . . . so now we are going to take these kids out? We didn't know how it would turn out."

Paasi and Rasmussen, on their way to a dry area in Chamber Seven, passed Karadzic and Brown. Their instructions were to tow their Skedco to a sandy spot in Chamber Seven and park there. But when they arrived at Chamber Seven something about the place felt wrong. The channel coursing beside the bank was still diveable—moving a boy through the water would certainly have been easier than hoofing him over the Martian terrain—so they wondered why they were there. Instinct told them to move; what they believed to be their mission orders dictated staying put—which they did.

A couple of hundred yards closer to the boys, Challen and Stanton took up a position at the beginning of the long submerged straightaway tunnel that connected Chamber Eight to Chamber Nine. Stanton's role, at least during the early phases of that first day, was to serve as "the feedback loop." Stanton was to wait for the first boy, assess his condition, and swim that information back to Harris in Chamber Nine. "Harry wanted to get feedback; were the boys alive or dead or heavily sedated, before he sent anyone else" down the tunnel, Stanton recalled. The three other British divers—Vollanthen, Jewell, and Mallinson—were already nearing Chamber Nine. They would each take a single boy, alone. Mallinson says (in his Yorkshire accent, which omits certain consonants, making "water" sound like "war") that "originally we had planned for two divers, one front and one back. Then we realized the conditions wouldn't make the diver at the back any use. He would've been trapped to the back [of the mini-procession]. He wouldn't be able to see what was going on even if that kid had panicked or his mask had flooded with water . . . and wouldn't have been able to help."

So there would be no help in the swimming sections along the route, only at the predetermined "gas stations," between which the divers would be entirely on their own. That meant about a thousand yards of gut-churning solo responsibility. There was almost no going back and no backup plan. If one of the boys' masks came loose or ripped and the divers happened to notice it while diving, they were instructed not to stop. They couldn't afford to fiddle with a mask once it had been carefully strapped on. And if one broke, there was only a single back-up mask that was bigger and older than the ones they were using. They just would have to make do.

"If one of the boys' face masks became filled with water," said Mallinson, "there was no plan B. You just got to swim him out as quickly as possible" to the next spot—always forward, never back—to find a spot that offered enough headroom to breathe and hope that enough oxygen from the tanks remained in the boy's system for a desperate resuscitation to succeed before he became brain dead or died.

Dr. Harris arrived at Chamber Nine first. In Harris's dry pouch was a note he'd dictated to a Thai doctor to translate; he handed the note to the SEALs, instructing them to read it verbatim to the boys. He needed to be able to manage the cave, so it was imperative that the SEALs followed these instructions. Dr. Bhak spoke to the kids, who seemed relieved. They could finally see the finish line, even if they wouldn't be conscious when crossing it. The SEALs gathered at the top of the mound in Chamber Nine and read them the instructions: They were to swallow a pill that would make them feel funny. They would come down to the water and sit on Dr. Harris's lap; he would give them an injection in one leg, then give them another injection in the other leg; they would fall asleep; they would wake up in the hospital.

Harris had administered ketamine and atropine plenty of times. And he had been in caves plenty of times, even at pulverizing depths. But, he thought to himself, he had never administered the sedative "in the back of a cave on malnourished, skinny, dehydrated

Thai kids before." Cool on the exterior, inside he was pulsating with fear.

Upon hearing the instructions, the boys didn't flinch.

Ignorance is bliss, Harris thought to himself. But it wasn't just ignorance. The boys knew it was dangerous. But they had been born and grew up in a society that inculcates almost blind obedience to authority. And not just to the government, or its king. At the Wat Doi Wao temple and monastery, where Coach Ek and the boys had spent so much time, leading monks are so revered that if one of them was present the boys would not walk up to him to pass a message or ask him something—they would crawl on all fours.

In the end, the boys sensed the danger, but they also trusted these authority figures instructing them on the way out of it. If they said this plan was going to work, then it would.

Chapter Eighteen
A Few Shots of Ketamine

M allinson had volunteered to go first. "I'm not one to hang around in the back. I'm pretty fast, so naturally I'm usually at the front, so I just volunteered." He knew that would mean a little less silt in his face on the way out, but likely also problems that no one else had encountered and no one had previously troubleshot.

He helped four boys into their wet suits, fit the inflatable vests on them, and readied an elastic band on each to help secure the oxygen tank. There were no formalities, and aside from cursory smiles, both the divers and the boys were all business. There may have been a reason the divers neglected to ask the boys' their names—a desire to build an emotional firewall between the grim job before them and the very real children who might die in their care. On that day the boys were just numbers in neoprene suits.

The SEALs led the first boy, Note, halfway down the slope, and the divers then led the woozy fourteen-year-old to the water's muddy edge. The rest of the boys were kept in the nook where they slept at the top of the slope so that they wouldn't freak out when they saw their friends anesthetized or bound. The Xanax had already pulled a gauzy veil of calm over Note's nervous system; it felt like being drunk without the mood alteration. The Brit and the Australian dug their feet in for a firmer purchase on the crumbling bank as the boy was handed over. Dr. Harris stood in chest-deep water, posting one leg higher on the slope and making a sort of shelf with it for the boy to perch on. He then jabbed a syringe into each leg.

Then there was the most troubling bit, the bit that the divers did not want the other boys to see. As Note faded from consciousness, Harris and Mallinson got to work, looping a zip tie around each wrist, then closing those loops with another tie and securing it with a carabiner. They had effectively handcuffed him. They also zip-tied the boy's legs together. Now came the part that every one of the divers had worried about: they tightly strapped the full face mask onto the boy's head. A leak could mean drowning. These were adult masks and—while some of the boys were tall—after twelve days of starvation their skin was so taut it looked as if it had been shrink-wrapped over their faces. The mask has five straps: one just above the forehead, two just above the ears, and another two where the jaw meets the ear. They yanked and they cinched.

"If they had been conscious it would have been very, very uncomfortable for them, because we really strapped it down tight," said Mallinson. For the next thirty seconds or so, said Harris, the first boy stopped breathing. Then his chest rose and he inhaled.

The oxygen tank now fully secure in the elastic around the boy's waist, he looked like a captured Martian. In Mallinson's hands was a package of flesh in the shape of a human—a creature in a state closer to coma than to sleep. When he settled into the water, Dr. Harris had to test the seal again, dunking the boy's head into the water.

Again, apnea: Note stopped breathing. An eternal thirty seconds ticked by before the bellows of the boy's diaphragm raised his belly and drew in air. Still alive . . . the bubbles bobbed reassuringly from the side of his mask.

The "package" was now in Mallinson's hands.

After nearly a quarter century of rescue diving, Mallinson says he usually doesn't get nervous—now he was nervous. He fiddled with the boy's buoyancy so that he wouldn't float to the surface or sink to the bottom, and watched the little black figure disappear into the water beneath him. Gripping the two straps on the back of the boy's inflatable vest and deflating his own vest, he started kicking. With decent visibility, he didn't have to hold the guideline, as long as he kept it in sight. Instead he put a second hand on those straps. The support team had rigged a tether to the inflatable vest, but Mallinson chose not to use it—he feared it could get tangled in the guideline or the cave's countless other snags. The first section of the dive back was the longest single dive of the extraction—about 350 yards. His eyes darted from the guideline to the release valves of the boy's mask. Note's breaths were slow and labored.

If stadium lights could have illuminated them in the water, the strange pair in black neoprene might have looked like equally sized sea lions on a leisurely swim, one directly above the other. Mallinson, at five foot five, was shorter than several of the boys whose lives he'd volunteered to save—and about the same height as Note. He decided to keep Note close, with his own head protruding just ahead of the boy's—so that his helmet would take the brunt of unseen stalactites.*

That first part of the tunnel narrows down to the size of a city sewer. It was a long, continuous swim of over twenty minutes. Mallinson refused to let his mind wander. He was thinking ahead to a nasty bit of choreography. Toward the end of the passage there is a choke point, which looks to a swimmer like a curtain of rock—a dead end. The guideline drops to the bottom of the canal, but there's only one way to fit through. Mallinson had to remember which side to put the boy on. If the boy stayed on his right, they'd get stuck. When they arrived, he tugged the boy to his left, so they were swimming abreast. He had to contort Note's body to get it through.

Note's head, facing down, inevitably clanged against the unseen rocks. If you float in a pool facedown, you'll notice that while it's easy to keep your abdomen floating just by breathing, your legs are not buoyant—they sink. Which is why the boy's bare feet dangled low

---

* The full face mask acted like a breathing tube, but it was so bulky it made putting a helmet on the boys impractical. There was also a concern that if the helmet knocked against a rock it would break the mask's seal.

and scraped the sharp rocks and gravel on the tunnel floor. But Mallinson wasn't worried about that. He also wasn't worried about the burst eardrums the kids might suffer in some of the deeper dives.* His mission, brutal as it sounds, wasn't necessarily to bring the boy out in one undamaged piece. It was just to bring him out alive. And for that, his sole focus became the seal on the mask.

"It felt as though it was quite firm on the face. But until you'd knock them about a bit—and I don't mean on purpose, it was just that you know you're in there with no visibility; you're going to hit a rock. Sometimes they got a bit of a bash. So I knew after the first five minutes that the mask had a good seal." His focus turned to those bubbles. When he couldn't see them he would pull the boy in, a finger's distance away from his face, hold his own breath and listen for the reassuring rattle of the boy's uneven breathing like a nervous mother peeking in on her sleeping newborn.

In Chamber Eight, Stanton felt the line vibrate. He watched the phantasmal couple emerge in a blast of headlights.

Stanton could not tell whether Note was breathing

---

* Unlike Mallinson and the other divers, the boys would not be able to equalize their air pressure.

or not. Mallinson handed him the boy, whom he and Challen put into recovery position on his side and began to examine. Mallinson waited a beat as he dragged himself out of the water—and said, "Oh, yeah, he's still breathing."

Challen tapped on the corners of the boy's open eyes—open eyes are part of the dissociative state under ketamine. Mallinson began to gather his cylinders to head for Chamber Seven, but he was irritated. He had just completed the arguably single most perilous leg of a rescue dive in history. Why the hell were Challen and Stanton alone? Where were the Euro-divers with the Skedco? After removing the boy's mask and checking on his breathing, pulse, and body temperature, Challen informed him there'd be no more help, no stretcher. So Mallinson and Stanton grabbed Mallinson's and Note's tanks and humped them to Chamber Seven. They returned for the boy. Challen took his arms as Stanton grabbed his legs, and they hoisted the limp body. They had about two hundred yards to cover. They kept tripping on the shark's-tooth rocks and strained to lift Note over and through a boulder field. The tunnel there meanders, and each section seems to have its own ecosystem of mud—ranging from crunchy and gravelly, to ankle-deep pudding, to beachy sand. It was tough going for the forty-nine-year-old Mallinson and the fifty-seven-year-old Stanton. When they reached the water on the other end they were heaving. Mallinson was exhausted.

The semisubmerged parts of the cave required more work to plow through than the totally submerged parts. The diver had to slosh through the water, bending low and dragging the boy on the surface. His hands ached from white-knuckling the straps in the hour since he'd left Chamber Nine. When Mallinson arrived at Paasi and Rasmussen's position, he barked, "What the hell are you doing here? You're not needed here—we need help back at Eight, you need to get back there . . . and bring the Skedco!" The two Euro-divers weren't quite sure what to say. They were there to assist and support—the Brits called the shots and there was little point explaining the various Talmudic interpretations of the definition of "Chamber Seven and Chamber Eight." They thought they *were* in the right place—but they had not been certain because they had never before been this far into the cave. What the plan had called for, but what had not apparently been specified clearly enough, is that Paasi and Rasmussen were to help the divers cross the dry areas of Pattaya Beach. As Paasi helped Mallinson move forward, the forty-five-year-old Rasmussen bolted back for the farther reaches of Pattaya Beach, alternately power-swimming and running, his Skedco bouncing over the rocks behind him.

He arrived to see Challen kneeling over the second boy, Tern, and Vollanthen unhooking his gear. Vollanthen glanced up, and for the second time in half an hour a Brit would bark at Rasmussen for being late or in the wrong place. These were tense moments, and any wrinkle in the plan might have had fatal consequences. Rasmussen and Challen pulled off the second boy's kit, laid him on the Skedco, and wrapped the firm plastic sides around him. They hustled him to Chamber Seven as Vollanthen carried the gear, jabbed him with a farewell dose of ketamine, and handed him to Vollanthen—who was already stationed in the water—and headed back. After an hour of continuous hustle it was a bit of a break. Rasmussen and Challen, who'd worked together on various cave-diving expeditions, chitchatted.

About twenty minutes later Chris Jewell arrived with the third boy, Nick, who had just had his fifteenth birthday in that rocky tomb. He wasn't the oldest boy, but he was one of the biggest, standing about five foot eight. Jewell was about fifteen years younger than the next-youngest Brit rescuer; as the Aussie and the Dane labored with Nick and the Skedco, they noticed Jewell had marched through the boggy mud without taking off his diving gear and carrying the boy's extra tank. It was well over a hundred pounds of gear and he wasn't out of breath. On their way back, as they meandered through the snaking tunnels to Chamber Eight, Rasmussen and Challen marveled at the younger man's strength. They were expecting another twenty-minute breather before the last diver of the day. While they had no idea whether the boys had survived the entirety of the journey, the Dane and the Aussie felt a flutter of optimism that this insane scheme might work.

About forty minutes earlier, swimming from Chamber Eight to Chamber Nine, Stanton thought the same. In that long underwater stretch he had just passed Vollanthen with Tern and seen those blessed little bubbles rise from the sides of the second boy's mask. When he swam up to Chamber Nine, Jewell was dressing his ward for the journey. Stanton informed Dr. Harris that two of the boys had survived the journey as far as the edge of Chamber Eight. He then helped Jewell and Harris pack the third boy off and prepared the fourth boy to be moved that day. It was Night. He was fourth because his cousin Nick lives about half a block farther from the cave site. Like his cousin, Night was one of the larger boys and seemed perfectly healthy coming down the slope.

Geography played a role in the boy's order of exit, but it so happened that the first four boys out were also the group's biggest—just as the American planners had hoped. It was no accident, but was a clever bit of decision-making by Coach Ek and Dr. Bhak, which they kept from the boys.

Night was one of the tallest boys, and while the foreign divers didn't know it yet, he'd also been one of the boys with the most severe symptoms of pneumonia. So as soon as Harris dosed him with ketamine, he stopped breathing. Temporary apnea under sedation is not abnormal and had happened with Note, as well. Once they fitted his mask and dunked his head in the water to ensure a seal, Night stopped breathing again. Then came a slow breath. Tentatively, Stanton grabbed hold of the boy, floating him above the water as he nosed into the canal. Ten yards, no breath; twenty yards, nothing; thirty yards, he couldn't tell. Fifty yards out, Stanton shouted back to Harris: "He doesn't seem to be breathing much!"

Dr. Harris shouted back, "There's nothing we can do about it, keep going!"

Over one thousand feet of rock and coffee-ground

water separated him from help. Stanton estimated Night was breathing about three times a minute. As he multitasked, he'd sometimes realize that a minute had gone by without his registering a breath from his charge.

Behind Stanton, Dr. Harris quickly packed up and set off. While the divers carrying the boys were still threading their way to freedom, Harris's work at Chamber Nine was done for the day—his end-of-day duty would be to serve as a rear guard, making sure that everyone got through okay. When Stanton pulled up to Chamber Eight, no one was there; he dragged the boy up the bank and waited. Within minutes Harris arrived, noticed the boy was starting to come to, and dispatched the boy back to narcosis land with another jab of ketamine. But again Night stopped breathing, this time for longer.

As Rasmussen and Challen rounded the last bend of the tunnel before it opened into Chamber Eight, they saw lights glancing off the walls. They heard men talking. At the straightaway thirty yards out, they saw Dr. Harris lying on the sand spooning Night, cradling his head, and trying to keep his airway open. They had no doubt that some of the boys would die, which explains why Rasmussen and Challen were not particularly surprised to encounter this tableau.

"He behaved like a kid with a bad chest infection under anesthesia, a lot of breath holding; he was oversedated," said Harris of Night, who had received two shots of ketamine less than an hour into his journey.* Harris lay there, cheek to sand, fearing the boy was slipping away and thinking, *Well, this is what I predicted would happen, this is going really badly.*

Ramussen went to see if he could help Harris, and Challen went to help Stanton gather his and the boy's gear for the hike to Chamber Seven. Dead or alive, Stanton would have to take this teenager out, so he thought there was little point in holding up the operation by waiting around here to watch this boy die. For a while no one could tell if Night was breathing. There was no death rattle, just sporadic sips of breath. He had been one of the thinner boys to begin with and was now even thinner. The neoprene suit was bulky and the boy's body seemed hummingbird frail—they had to bend low, in silence, to detect a sign of life. On his knees, Rasmussen—who speaks rudimentary Thai—leaned in close, held the boy's face in his hands, and began cooing softly to him with the patchwork phrases he used when talking to his four-year-old Thai-speaking daughter: "Don't worry, son—you're on

---

* Harris said this at the Swan Trauma 2018 Conference.

your way home—you're going to your mom. . . . Don't worry, son—you're on your way home—you're going to your mom." It became a mantra Rasmussen would repeat over and over for the next thirty minutes, partly to soothe the boy and partly to calm his own nerves. From then on, he would repeat those words to each of the boys who came through Chamber Eight.*

After half an hour "the boy sort of fired up, and we ended up needing another dose as we put him in the water down the track (to the sump at the next chamber)," said Harris. Stanton was the last rescue diver, and the previous divers had kicked up so much silt that he was blinded. Visibility was now barely a foot, and at times he could not make out the dainty little bubbles floating up from the boy's mask. So, as he inched along the tunnels, instead of holding Night by the straps on his inflatable vest Stanton wrapped his arm around

---

* Rasmussen likely knew this only intuitively, but studies have shown that patients in narcosis—or even in comas—can hear. Hearing is the last sense to go when the body shuts down, whether from induced narcosis, injury, illness, or medical procedure. And while it may only be a one-way conversation, those who have emerged from these near-vegetative states say they definitely heard conversation. In addition, ketamine sedation is sometimes used for auditory tests on young children who might not be able to communicate or are too jumpy for conventional tests.

the boy, holding him tight to his body and tucking the boy's head right under his chin.

By now Mallinson, with Note in tow, had transited out of Chamber Seven when he felt a disquieting twitch. His package was moving. The boy was coming to! Mallinson was now in neck-deep water. His mind was whirring. He had to find shallower water, but there was no time; his young ward was clearly waking up. He had to sedate him, and immediately. First he had to try to get Note, who was floating horizontally, into a vertical position. He wrestled him upward, his bobbing head now pointing to the cave ceiling. Using his left hand and knee, he pinned Note against the cave wall. He then used his right hand to grasp one of the boy's legs, which he hoisted above the surface of the water. Mallinson was already breathing heavily.

With one hand and knee holding the boy in position, thigh out of the water, he used his right hand to fish around in his dry bag for the syringe pack. But when he reached in and opened it, the syringes and needles popped out. There was suddenly a flotsam of syringes and needles bobbing in the water, slowly being carried away from him by the current. Still holding on to the boy, he started swiping at the wayward syringes with his free hand. Finally he was able to grab a syringe, attach a needle, hoist the boy's thigh higher out of the water, aim for the center mass of muscle, and jam it in. Then he waited, the two of them leaning against the cave wall, Mallinson panting. The twitching subsided, and Note was again comatose. It was the first time Mallinson had ever administered an injection to a human. He would give the boy three more injections before they reached Chamber Three.

In the brown gloom, Karadzic and Brown waited silently in Chamber Five, staring in the direction of the T-junction from where the boy and his handler would be coming. It had been nearly two hours, when Karadzic saw the water stir and the lights dance on the cave ceiling. In his dry suit, the forty-five-year-old suddenly sprang to alertness from his stupor and raced the fifty yards through waist-high water to Mallinson and the boy. As he splashed over, he called out, "Do you need help?" The answer was an emphatic *yes.*

As he neared Mallinson, he misread the distress on his face. Everything was fine, but after nearly one thousand yards of diving, swimming, and frantic sedating, Mallinson was exhausted. It was about three hundred yards to the next sump, where Mallinson could again dive. They inflated the boy's vest so he would float, and Karadzic took control, guiding the inert boy to Brown and their little way station. Karadzic's fingers were trembling, but he quickly switched out the boy's oxygen tank for a fresh one. As he took his pulse and measured his temperature, he noted that the boy was mumbling something in Thai. "I realized the boys weren't as sedated as they should have been." That was not a function of the medical care, but of the drug itself, with its short duration of action.

The last choke point in the dive was that drainage-pipe-size squeeze leading from Chamber Four to Chamber Three. Swimming out from Chamber Three to Chamber Four, a diver wouldn't notice the difficulty at the end. But on the way back into Chamber Three, it looked like there was a rock curtain hanging from the cave roof. Unlike the tunnel beyond, this was a vertical squeeze before the dip down into that choke point. Divers could only fit through one at a time—so each rescuer would have to slide the boy or himself in first. It was another landmark that Mallinson had committed to memory. And by the time he swam up to it, he'd already mapped it out. He rearranged the position of the boy, pulling him upright—his head sagging to his chest, legs still dangling in the water. Then Mallinson stuffed him through with his right arm and slid in behind him, careful never to let go of his precious cargo.

Because the traffic through this section had kicked up so much silt, it was one of the darkest parts of the dive, and Mallinson hoped his banged-up ward was still alive. Ahead was that home stretch of 150 yards to Chamber Three, where there were dozens of rescuers eager to help. Note was homeward bound; all that remained was the obstacle course of short swims, ropes, and a long, wet walk through chest-deep water.

A team of Americans, including a USAF Special Tactics pararescuer, and the Thai Navy SEALs waited at the sump of Chamber Three. There is nothing grand about the place or the sump—it is simply a large pothole in the floor of an unimpressive room the height of an average American.

It had been about six hours since the first divers slid into the hole. For the past hour or so, every eyeball in Chamber Three had been focused on the guideline in the water. There were no communications beyond that sump, no way to know whether a diver needed help or, if so, what kind. And no way to know if their precious cargo was still alive. Even if they became aware of trouble, it had already been decided that sending emergency divers to the rescue could trigger a potentially deadly collision or—nearly as dangerous—a human logjam in cramped spaces. Like a fish on a hook, they would only know they had something when the line started to bob. The more intense the vibrations, the closer the diver. Around 4 P.M. the line started to wiggle. Then it wiggled more vigorously. Finally two heads broke the mirrorlike surface nearly simultaneously. One of them was Mallinson's.

The U.S. Special Tactics team's pararescuer, Tech Sergeant Ken O'Brien, pulled the boy from the exhausted diver, raised him up out of the water, and flipped his limp body onto its side. No one spoke as Sergeant O'Brien bent low to put his ear to the boy's mask. The "package" was then strapped into a Skedco, which the Americans harnessed to the rope and pulley system and lifted gently over a series of boulders. After that the litter was carried by another team for about seventy yards around stalagmites and boulders to another rope system, and attached to another rope line. Thai SEALs then maneuvered the stretcher down the forty-five-degree slope to an American pararescuer, who would dive it to Chamber Two.

Mario Wild from the Chiang Mai climbing team was waiting atop the sump at Chamber Two, keeping the rope taut. A series of ancient military-style phones connected the first three chambers of the cave to the outside world.* They had to be cranked by hand in

---

* The Israeli phone system was returned to the company after the boys were found on July 2.

order to make a call, but the old technology was cave-proof. The phone buzzed in Chamber Two. It was a terse message informing the Thai commander there that the Skedco was on its way.

Said Wild, "We didn't know whether the kid was alive or dead. He had a full face mask on, couldn't see anything. It wasn't our job to assess whether the kid was alive or not. Our job was to get the kid out of the water and pass him on to the doctors."

Note was unresponsive, and it took a few seconds for the Thai doctor to assess his vitals. There was silence as the medic listened for breathing.

"He's alive!" came the call. Even though members of the teams were instructed to stay planted at their stations, there was a burst of cheers and a rush to take pictures of this marvel cocooned in plastic.

"It was an amazing feeling. It was amazing," remembered Wild, laughing. "The kid was alive! It took me a long time to process what had happened. Weeks, actually."

Mallinson was there and knew, of course, that the boy was still alive. The hardest part was over, but Mallinson wanted to see it through. He walked along as the boy on the stretcher was passed hand to hand and clipped in and out of various rope systems. Note was finally handed to a Thai SEAL team, which hauled him through more than four hundred yards of chest-high water and then ran him to the cave entrance, where he was exposed to his first rays of natural light in over two weeks.

Shortly after 4 P.M., the WhatsApp icon on Major Hodges's phone flashed red with a message from his captain, Mitch Torrel: "Hey, kid's out and he's safe."

With Anderson in charge of the granular planning, Hodges's role was leading the mission and serving as its liaison to the U.S. command structure and the Thai government. He couldn't afford misinformation. So just as he had when the boys were found, he asked for confirmation:

"Are you sure? Are you absolutely sure? The kid's breathing?"

Response: "Yup. Kid is doin' fine. He's laying here. And we're working to get him from the third chamber back to the second."

Hodges sat back. "It was huge," he remembered. "Wow, our plan is actually working. We can do this. We knew once they got to the third chamber—it wasn't as if they were home free at that point. But the hard part was done."

Note, who was the mission's unwitting guinea pig,

was definitely not home free. It took another hour for him to be hustled out of the cave, wrapped in the familiar package: full face mask, oxygen tank, sixteen-year-old boy, all enfolded in that plastic sled stretcher, with Mallinson carefully watching the whole way. Anderson decided to walk down the slope from headquarters and watch history being borne out of that wretched cave.

At its mouth he met Mallinson, who told him, "Derek, they did a tank swap at [Chamber] Three, but you might want to check that gauge." What Mallinson was getting at was that the boy had been on the same oxygen tank for about an hour. Those positive-pressure full face masks served the boys well in the water—pumping oxygen-rich air into their systems. However, if the air in the tank ran out the mask would become suffocating—like having a pillow pressed down on your face. Drawing a breath would require sucking air in hard through the rubber gauge at the sides of the mask. A focused and conscious person could do that, but not a comatose child.

Anderson checked the tank's pressure gauge, and his eyes bugged when he saw the needle was deep into the red, all the way at the last tick of the dial. The kid was minutes away from suffocating.

Anderson grabbed the doctor accompanying the

stretcher. "Hey, either this mask comes off right now, or he has to be up at this field hospital within the next five minutes."

The incident was sobering for Anderson: he now understood that even though the most perilous part of the rescue was the solo effort from Chamber Nine to Chamber Three, risk continued to accompany the boys all the way out. He realized that, given the complexity of the mission, until they were actually conscious and stabilized in the field hospital set up at camp they were not out of harm's way.

The next two boys, Tern and Nick, came out without any problems; Stanton, meanwhile, was bringing up the rear with the pneumonic boy—Night, the boy Dr. Harris expected to lose after that close call at Chamber Eight. At Chamber Five, alongside Karadzic and Brown, Stanton gave the boy another shot—his fourth. With Night comatose again, Stanton gathered himself for the trickiest part of the dive just after Chamber Four—the vertical line trap just before the last stretch of the dive—the one all the divers considered the most challenging. At this spot the guideline shot straight up and divers had to memorize both its location and the choreography of slipping through. Visibility was six inches, so Stanton had no landmarks to indicate he was nearing the trap. He had been caught in that spot on previous dives. "You can feel on the line that it's going to somewhere impossible. It's hard to find the right slot to get through."

He described it as picking your way blindfolded through a large room that has furniture strewn everywhere. There is a rope through the room, but it's entangled in the stubby legs of a couch, or around the upended back of a chair. You have to feel your way around those dead ends while never letting go of the rope. Feeling one's way is made more difficult when one doesn't have a hand to spare—he was holding both the line and the boy. Stanton knew he was tethered to the boy, and at times he would simply place him low on the floor of the tunnel, resting on his oxygen cylinder, as the diver rummaged around to find the exit. Finally he found it, stuffed Night through, and steadily kicked the last 150 yards to Chamber Three.

The kid sure felt lifeless in Stanton's hands. There was not a single discernible sign of life from Night. The line shook, and U.S. pararescueman Ken O'Brien called out, "Fish on!" Stanton pushed Night out ahead of him. As O'Brien hauled him out, Stanton shouted, "Is he alive!?" The chamber went silent once again as O'Brien placed his ear to the boy's chest, listened for a moment, then pumped out a thumbs-up: "He's alive!" After the hour-long obstacle course with the ropes and the stretcher relay, he would at least be in an ambulance, if not necessarily home free.

Even when the boys reached the ambulances, Anderson reminded the medics to maintain vigilance. The equipment on the comatose figures had to be carefully removed, so that it could be used again. There weren't enough suits for all the boys, and only four masks. Over the previous few days, donors from dive shops and rescue organizations from around the world sent hundreds of full face masks—normally costing about eight hundred dollars each—to the camp. Not one of them was a positive-pressure mask. So, among the thousands of rescuers and volunteers and the millions of dollars' worth of equipment stockpiled, there were only four masks that they knew would work—all of them brought by the U.S. Special Ops team. That meant they had to be gingerly removed once the boys were taken out of those flexible stretchers and loaded into an ambulance. More than once, Anderson recalls, he had to grab a doctor's arm to prevent him from cutting into a suit or a mask's rubber straps.

I was preparing to go live for *Good Morning America* at the crossroads about four hundred yards from the mouth of the cave when emergency lights from a vehicle began reflecting off the trees in the twilight. *Police or an ambulance?* we wondered. Moments later we saw it, rolling unhurried down the rutted road—there were lights, and then a brief scream of sirens. Over the hours we had been providing wall-to-wall coverage, there had been very little communication with the media. We had watched ambulances come and go, so we knew something was happening when police and soldiers corralled the press in a pineapple field. We knew the rescue had started and had been alerted it would take eight to ten hours, but were ignorant of its outcome. There were cheers as the ambulance sped by; most of the journalists and rubberneckers who'd come out to watch were reflexively elated, but there was no way to tell if the ambulance was serving as a mobile doctor's office or a hearse. But pretty soon someone believed to be in Chamber Three began leaking information to the media. There were four boys, they were all out, they were all alive. It would take days for the world to learn about the use of sedation and some of the close calls.

And while the world was just waking up to the incredible feat that these divers had accomplished, the rescuers were busy preparing to dive four more boys to safety.

Chapter Nineteen
The Complacency Gap

The system was untried, unorthodox, and yet somehow successful. Still, there was room for improvement.

Mallinson had been the first diver out, but he didn't stop working. While the rest of his team had been threading its way through the tunnel with the last of the boys, he worked on preparing gear for the next day's dive—tanks, regulators, and the boys' gear. None of them could stop that night—there was homework to do. The dozens of empty tanks had to be refilled with air and replaced, the ropes had to be tensioned.

And there were other systemic issues to talk about. Mallinson told Anderson they needed a stretcher at the dry area after Chamber Eight and another person to help Dr. Challen. Two men simply could not haul the boys in the alien dark for hundreds of yards—it was dangerously exhausting. Anderson had to recalculate the number of oxygen tanks for the boys—they clearly needed more of them between Chamber Three and the mouth of the cave. And there was something more important, Anderson told the group, which included the Thai Navy SEALs, the Chinese team, and the Chiang Mai climbing team: "Everybody needs to take a step back, relax, take a deep breath. Once [the boys] come out, they're still alive and breathing, you have to change their tanks. We'll systematically assess them, put blankets on them. There's no reason to be panicking."

He was specifically referring to some of the rescuers in Chambers Two and Three who'd teetered on the edge of nervous breakdown each time a boy was brought through, then crowded around as if to verify that the boys were breathing and snapped pictures. From now on, Anderson informed them, no one was allowed to leave their station until the last boys were out. And there was one more thing: they needed to know the kids' names, needed to account for who was leaving. They decided the divers needed to talk to the boys and write nicknames on their wrists with indelible marker.

There were also challenges for the newly deputized diver anesthesiologists. That night, Sunday, July 8,

Karadzic felt ill. He said he had a fever and the camp doctor benched him. Given that everybody was burnt out already, Stanton's take was that the real source of the mystery illness was Karadzic's understandable discomfort with administering sedatives to mumbling, nearly comatose kids. He lived and worked in Thailand, and knew what a mistake would likely cost him. Regardless of whether it was because of that or because he really was sick, Karadzic sat out the next two days.

Adding to the stress of administering the ketamine were the dose sizes themselves. The divers were given a smaller dose for the smaller boys and a bigger dose for the bigger ones. But with the kids in wet suits, their hands bound, it was hard to tell who was big and who was little, particularly with the thirteen-to- fourteen-year-olds. It made for some agonizing decision-making and fumbling for syringes in the photon-free dark. Most understood enough to know that what Dr. Harris had told them about their inability to overdose the boys was not entirely accurate when dealing with emaciated children. So the team asked Harris to just provide them with a single generic dose to simplify the decision-making process.

Whether it was because of these adjustments or because the team had already been through the process once, the most memorable thing about the next round of rescues on the following day, Monday, July 9, was that none of the divers remembers much about it. It went remarkably smoothly. Harris told the remaining eight boys and their coach that their rescued teammates were safe and recovering in a hospital. As soon as the boys padded down the slope to him at the waterline with the help of the SEALs, Harris began chirping at them in his Aussie accent, flashing his gap-toothed grin, knowing full well that they wouldn't understand him. His goal was keeping things light in that creepy place. There's a term medics use for it: psychological first aid. So Dr. Harris deployed his Wet Mules brand of humor. One time, as he buried one needle, then the other, into a boy's legs, he quipped to no one in particular, "Oy betcha didn't see THET one coming, then."

They didn't. The shot of ketamine was nearly painless, but the boys seemed to suffer more with the atropine, the drug that would ensure they didn't choke on their own saliva or mucus. They winced, they groaned, but not a single boy cried.

On that second day, Nick, Adul, Biw, and Dom were brought out without a single incident or scare. The ambulances crunched over the gravel, moving away from the Tham Luang cave bound for the soccer field where the boys had last scrimmaged two weeks earlier.

Their pitch had been turned into a helicopter landing zone, now home to two Russian-built Mi-17 transport helicopters. Within minutes they were in a hospital in Chiang Rai. It seemed breathtakingly simple.

On their Facebook page that evening, the Thai Navy SEALs posted "8 boys out in 2 days—Hooyah! (The unit's rallying cry)." It was promptly commented on by Mark Zuckerberg himself: "From everyone at Facebook—your bravery has been amazing and congratulations on the successful rescue of eight Wild Boars. Best of luck as you work to get the remaining four players and their coach to safety." Eight dives, eight boys, eight successful rescues. The rescuers, who had initially calculated only one of those original eight would survive, were now eight for eight.

About ninety minutes after Dom, the eighth boy rescued, was pulled out that day, Thailand's prime minister and junta leader, retired general Prayut Chan-o-cha, rumbled to the site in his motorcade. First, the prime minister, followed by his entourage, marched to the SEAL tents beside the cave. The old infantryman led the SEALs in a few rounds of "Hooyah!" Watching from a distance was American climber Josh Morris—who was now acting as de facto liaison between the Thai government and the foreign divers after his deft handling of that climactic meeting with the interior minister two days prior. He would be assigned to translate for the prime minister when he later met the foreign divers, but as he watched the prime minister he thought to himself, *These men needed this boost. What a great thing for them.*

After meeting with the SEALs, the prime minister followed Morris and others, making a beeline for the four-pole tent where the British team was tucking into some of the noodles and green papaya salads prepared by the camp kitchens. The man who had taken power in a coup in 2014 arrived holding a large box of snacks for the divers, and with an arched eyebrow began reading the labels: "Ah, these are new coconut snacks in Thailand."

That broke the ice.

There was little formality, aside from a few handshakes and a couple of snapshots. The prime minister gamely took a seat on one of the plastic chairs encrusted with a film of cave junk; surveying the table, he congratulated them on the eight successful rescues, but added that the job was not complete, so "no drinking tonight, no drinking yet!"

The prime minister then asked if there was anything he could do for them.

"Yeah," said Stanton, glancing at Vern, who had to leave the country after each sixty-day visit on his tourist visa. "Get Vern a [permanent] visa!" They all had a laugh at his cheekiness.

By then the prime minister had been introduced to every member of the group—including Vern's life partner, Tik—and so he quipped, "Vern, why don't you just marry Tik and then you can have a visa." He then reached into his breast pocket for a pen, suggesting that he would officiate the union right there. Vern called out, "No, not yet!" Now they roared with laughter.

By now the rescuers were feeling really good and quite comfortable with the prime minister; Vollanthen pointed at the two Tourist Police officers who had been assigned to be their minders-cum-shadows. "These two"—he jabbed a finger in the direction of the uniformed men, whose eyes so widened in anticipatory terror that Stanton was worried "they might shit themselves." "These two," Vollanthen went on, "have been excellent to us, they deserve recognition." Again, peals of self-congratulatory laughter. The two lead British divers had become quite close with their escorts, who also served as their aides-de-camp, even helping prepare their gear and stuffing those MREs into the dry bags for the boys a few days earlier. As the prime minister left, he gave Vern a friendly smack on the belly and a wink, as if to say, "Marry the girl," and walked off into the night.

The scene was lighthearted and congratulatory, which made Hodges and Anderson all the more nervous. In the military, said Anderson, "we always reminded ourselves that after a success, there's a huge gap for complacency. And complacency always works its way in."

Just after the completion of that second day's rescues, Anderson was chatting with Harris. "Doc, you know, what are your thoughts? Eight are out. There are five left."

"Man, this has never been done before," Harris answered. "This is actually working. We're actually succeeding at mission impossible." And then he added, "I was fighting the urge to get a little prideful." Harris possessed the self-awareness to detect the blaring alarm of complacency and to know that it often leads to failure. Anderson wanted to ensure that everyone was aware of it, so he gathered the team that night and said: "Hey, we are not done until all thirteen are outta the cave, and all four navy SEALs are out."

He had reason to be nervous. The forecast overnight—ahead of the third day of rescues—was for more rain, possibly a couple of inches. The U.S.

planners had decided they would suspend the next day's rescue if the rainfall topped fifteen millimeters (about two-thirds of an inch) an hour. Anything above that mark would likely overwhelm the pumps inside the cave and Thanet's dam above the Monk's Series. And if that happened there was no telling how long they might have to leave the remaining boys, the coach, and the four SEALs in there. That fear was compounded by the understanding that the rescue divers could not camp outside the cave indefinitely: Vollanthen and Stanton had already been there nearly two weeks, and they and the other eleven rescue divers eventually had to return to their jobs and families. But Anderson needn't have said anything. Many of the divers wore good humor as a mask hiding their own private worries.

"Now the tables had turned," said Chris Jewell. "Now, actually, the expectation was that we would get them all out. So suddenly the chance of losing a single child would be catastrophic, and the pressure on us individually increased at that point."

It was a stunning turnabout from forty-eight hours earlier, when they'd thought getting just a couple of the boys out alive would be a "success." They'd become victims of their own skill, with the increased weight of expectation upon them. Now Stanton started wondering what would happen if they lost a single child. "There was so much pressure after that second day. Everyone's expectations reset for 100-percent success. If there had been [one who died], would the mission have been seen as a failure?"

The Thai prime minister actually had one other person to meet that night, the tech titan and founder of Tesla, SpaceX, and The Boring Company, Elon Musk. Six days earlier, on the afternoon of Tuesday, July 3, a Twitter user based in Swaziland, a tiny African nation squashed between South Africa and Mozambique, politely tweeted at Elon Musk:

> "Hi sir, if possible can you assist in any way to get the 12 Thailand boys and their coach out of the cave. @ElonMusk."

The Twitter user, going by the handle @Mabz, wasn't particularly active on the social network and had about five hundred Twitter followers at the time; Musk had something on the order of 22 million. Luckily for Mabz, whose real name is Mabuya Magagula, a software developer for the Swaziland Revenue Service, Musk resides in California but seems to live on Twitter. That among the meteor shower of tweets bursting onto his page every day he spotted this one, from a complete unknown, and decided to respond to it, is astounding. Magagula wasn't so surprised that he had responded, but "I was really surprised that Musk put himself on the line like that. Others would have decided to avoid [involvement in the rescue] because it could have ended badly."

Magagula also knew that the polymath tech titan has a reputation for grand gestures of largesse. A day later, on July 4, Musk replied, "I suspect the Thai govt has this under control, but I'm happy to help if there is a way to do so." That response generated more than twenty thousand likes. Magagula is the father of a one-year-old, and at night in his comfortable bed he couldn't shake the image of the boys in the cave: hungry, tired, yearning for sunlight and their parents. He was glad Musk felt the same way he did: "I was really impressed with his humanity."

The tech maestro then ruminated on it for a day or two, and on July 6 retweeted an idea floated by the CEO of a satellite communications provider to create an air-inflated nylon tube through parts of the cave that would work like the guts of a bounce house, allowing the boys to scramble out without getting wet.

Musk's chief of staff reached out to Rick Stanton,

and that generated calls and an e-mail exchange between Stanton and Musk himself early on July 7—all this while the team was in the midst of those practice dives with the local boys and the ROC drill.

From: Elon Musk
To: Rick Stanton

Are you one of the divers who understands the cave geometry? Trying to help out, but need to know the details of the most problematic areas.

Stanton explained to him the tortured contours of the cave—the daggerlike rocks, the spaces smaller than the crawl area beneath a kitchen table, and the appalling visibility. One of the world's most innovative engineers was on the case, and he was indeed talking to cave-diving experts involved in the rescue itself. Musk now appeared intrigued, and later that morning of Saturday, July 7, rifled off an excited tweet about his brainstorming session—he called it "iterating" with "cave-diving experts in Thailand"—about creating "an escape pod" that would be alternately pushed and pulled out of the cave by divers. In that same tweet he also floated the idea of the inflatable tube, but noted it was "less likely to work given tricky contours of the cave." That tweet was "liked" over 150,000 times, or by nearly twice the number of people who live in Swaziland's capital of Mbabane.

Musk assigned SpaceX rocket scientists and space-suit engineers to the case and threw his tremendous energy into the project, exchanging multiple e-mails over the course of that Saturday with the terse Stanton, who at one point wrote back: "With respect all I see is a tube, albeit made of fancy materials." Stanton noted that they had been ferrying food using something simi-lar to those dry tubes (which is like comparing a push-cart to a Tesla), and asked for specifics about breathing systems, venting systems, and buoyancy control.

Early on Sunday, July 8, Musk, who was starting from scratch and investing tremendous resources in the project and who had been remarkably cordial, was of-fended: "With respect, I am trying to be helpful. Please do not be rude."

Musk was right, and Stanton apologized, encourag-ing the inventor to push on. At that point, even though the Thai prime minister had approved the divers' res-cue plan and it was scheduled to start in hours, the rescue team had no plan B. If they couldn't dive the boys out they would die, so Stanton was eager for any mechanism that might protect them; so far, Musk's idea had been the only viable alternative. "We're worried about the smallest lad please keep working on the capsule [escape pod] details," Stanton wrote in one of several e-mails that Musk later posted on Twitter.

Musk asked for more feedback, adding that he didn't want to ship the cigar-shaped tube to Thailand if it was not going to be useful. His SpaceX engineers had swiped a nearly six-foot-long component that funnels liquid oxygen into the Falcon 9 rocket's engine. The tubes were extraordinarily robust and made out of aluminum lithium alloy. As Musk assigned some of his rocket people to shape that pipe into a cigar-shaped escape pod, he also sent engineers from The Boring Company to the Tham Luang cave to assess whether their technology might help drill a shaft to the boys.

Musk's next series of excited tweets, late on July 8, announced the rough specs: "primary path is basically a tiny, kid-size submarine using the liquid oxygen transfer tube of Falcon rocket as hull. Light enough to be carried by 2 divers, small enough to get through narrow gaps. Extremely robust." The principle was basically this: instead of maneuvering the boys out of the cave with an air tank strapped to them, they would be maneuvered *inside* an air tank, cocooned from the harsh elements of the cave.

Later that day he tweeted videos of his pod, noting

that it would likely be ready to ship to Thailand in eight hours, and that seventeen hours later—late on Monday, July 9—it would be in Chiang Rai. The prototype that the SpaceX engineers built and tested in Los Angeles pools, and which Musk proudly displayed on Twitter and Instagram, looked like a gleaming six-foot-long silver torpedo with a hatch on its flat bottom. That's where the "casualty" would be placed, whereupon a team of two divers, pulling and pushing, would theoretically guide the space-proof capsule through the midnight murk of the cave's tunnels. That bottom hatch featured a mechanism which pumped air into the pod from what appeared to be air tanks strapped to the torpedo's belly. It was just wide enough for a skinny adult to squeeze in, but big enough for the Thai youth soccer players.

In the last video Musk posted of the pod, a group of divers navigate it to the edge of a pool where a team is waiting with monkey wrenches. The video shows at least five people on the pool deck pulling it out of the water, with two others pushing from the pool, and within forty seconds or so the "casualty" wiggles out. Though bulky and heavy, it had a lovely design—truly a Musk-esque craft, created by some of the most brilliant engineers in the world at breakneck speed. And Musk promised that it would also be safe, writing that the "operating principle is same as spacecraft design—no loss of life even with two failures."

For Stanton, some of the specifics were left wanting; for instance, what would happen to the carbon dioxide building up inside the tube as the boys exhaled over a two-hour period? The man in the pod had a pony tank with him, a small scuba tank fitted directly to a mouthpiece—would a rescuee need that as well? The difference between Musk's plan and the one devised by British divers and the international team could not have been greater. Ex nihilo, Musk's paean to engineering was a sleek silver pod made from futuristic rocket parts. The British plan could have been distilled into "drug 'em 'n' drag 'em."

Musk's tube appeared to be too big and too late. By the time Musk landed in Thailand on Sunday night, July 9, the divers had already rescued eight of the boys. Musk wrote courteously to Stanton: "Have to leave for Shanghai tomorrow morning, so don't know if we will have the opportunity to meet. If not, please allow me to express my admiration for the incredible job that you and the rest of the dive team have done."

By then, Governor Narongsak had publicly thanked Musk, but called the device "impractical." It was a bit of a head-scratcher for the media, because at this point the rescue of the remaining four boys and their coach

364 • MATT GUTMAN

seemed a mere formality. "It was a very well-meaning effort" is how a U.S. official diplomatically described Musk's team's initiative, "but for the conditions in the cave it wasn't going to work. The [British] divers were pretty clear with him that it was not viable for this one, but could be used for other situations," the official added.

On the night of July 9, Musk met the prime minister. According to diplomatic sources, during this meeting Musk somehow understood that the prime minister had granted him permission to enter the cave. So, in the middle of the night, with operations for the third day due to begin within hours, Musk managed to breach the careful order that the American team had implemented and, according to the sources, bluster his way in.*

In his time there he managed to snap a couple of stunning nighttime shots and some of the clearest video yet of what appears to be the swampy canal leading out of Chamber One toward Chamber Two, which he mistakenly called "Cave 3," and posted them to social media. He was at the nerve center of the rescue hours

---

* Musk's representatives say he was "encouraged by army and navy leaders on site" to go in.

before it was to resume. Thoroughly unimpressed with Musk's presence were the Thai leadership at the site, including Governor Narongsak, the King's Guard, and other honchos.

"They were pretty upset," said the diplomatic source, and the U.S. State Department began to brace for blowback. Aside from the dismay on the Thai side, there was no fallout, and Musk graciously left his escape pod with the Thai Royal Navy.

The last e-mail in the Stanton-Musk chain consisted of a query by Stanton about why Musk's engineers hadn't brought the tube up to the camp to present it to the actual rescue divers—after all, it might be used in future rescues. Stanton says he never got a response. The rescue divers suspected the engineers might not have wanted them to inspect it too closely. However, others at the camp got to see it up close, including Thanet.

Chapter Twenty
Lost

It had rained overnight—hard. When Thanet Na-tisri, the water-management consultant, woke up on the morning of Tuesday, July 10, it was still pissing down.

The rescuers had originally hoped that this would be a down day. Stanton and Vollanthen had been going nonstop for fifteen consecutive days. The rescues were brutal on the body and the mind. Television networks were going with wall-to-wall coverage of the rescues—so many millions watched that it was sapping viewership of the World Cup, and Elon Musk's arrival only magnified the attention.

But the rain forced their hand. Looking at the forecast, the rescuers determined that the weather would likely only get worse, so there would be no day off on

Tuesday. In fact, the rescuers were so concerned about the weather that they decided to extract the last four boys and the coach all in one day, before conditions deteriorated further. This decision to remove all five in one day had its own complications, because they only had four positive-pressure masks that they knew would fit the boys. There was a fifth, but it was an older model designed for a slightly wider face. Well over one hundred full face masks had been flown in from around the world, including some designed for women and children, but none had the special gauge that would turn them into positive-pressure masks. And by the time they got the go-ahead for the mission, planners deemed it too late to scrounge up another, which would likely have to be flown in from the United States.

In addition to the problem of the masks, there was also the issue of divers. Mallinson, Vollathen, Jewell, and Stanton were the divers best qualified to bring the boys out, and they had taken one kid apiece on each of the first two days of the rescue. Now, with five people to take out on this final day, something had to change. Mallinson, arguably the strongest swimmer and as experienced as Stanton and Vollanthen, volunteered for double duty. He would take the first boy to Chamber Seven, where support diver Jim Warny would take over and haul the boy to Chamber Three. Mallinson would then swim back to Chamber Nine and take the last person—they assumed it would be the coach—to freedom.

With the day's rescue of five people now hanging in the balance, Thanet scrambled to figure out how much rain had accumulated so far. A decision had been made that if it rained fifteen millimeters—a little over half an inch—an hour that day's operation would be canceled. Thanet called his weather team at Thailand's Geospacial Engineering and Innovation Center, then dialed the workers maintaining the dam above the Monk's Series, which likely had a hand in diverting some of the water flow inundating the cave. For nearly a week, the weather, along with Thanet's water diversions atop the Monk's Series and the pumps the Thai government had set up inside the cave, had allowed for more favorable diving conditions. Those factors had been partially responsible for the success they'd had on the first two days; after all, had the water level in the cave not been reduced, the divers would never have been able to reach the boys in the first place.

And now all that appeared to be shifting. The element Governor Narongsak had declared "the enemy of the rescue"—water—was again at the gates. Inside the cave, conditions were worsening—thirteen divers flapping around over two days was more activity in the deeper recesses of the tunnel than it had experienced at any point by an order of magnitude. The divers now faced silt that threatened to clog rebreathers and mouthpieces, and now a deluge of rain foreboded a stiffer current. That they foresaw this didn't make it easier. This is why the Americans had pressed for extraction as soon as possible—and why they scrambled to start that day's rescues more than an hour ahead of schedule.

When Thanet spoke to his team working the dam above the Monk's Series, he was told it had rained fourteen millimeters in an hour. He called Anderson with the news just before the dive teams were about to set off. With rainfall just one millimeter—the width of a dime—short of the predetermined cutoff it was a tough call; of course, once the divers dipped into the chilly waters of Chamber Three, it would be impossible to pull them back. Anderson was standing by on his WhatsApp ready to tap out an "abort!" message to Chamber Three if the rain increased.

For their part, the divers had their own barometer, and it wasn't Thanet's rain gauges. "I was ready to abort at any minute," said Stanton. He could plainly see it was the most significant precipitation since he had watched the cave flood ten days earlier. Around Mae Sai that morning, the grass got soggy, then disappeared into bog. The streets were slicked at first, then little canals started forming at the sides. It rained for hours.

And much more rain was coming—it was perhaps now or never. A delay today could leave the last five members of the soccer team and those four Thai SEALs marooned for months, possibly until death. Hodges and Anderson decided to proceed. The mission was still a go.

The one reason for optimism was that the day's actual forecast, if it could be believed, was for a break in the rain from late morning into afternoon, before another system was predicted to blow in. That small break in the precipitation might give them the time they needed for a successful mission.

Stanton was prepared to make the call once they got to the T-junction. If the team noticed clearer, warmer water rushing in from the Monk's Series on the right, they would turn back. It would also be the ultimate test of the dam and diversion of that unnamed stream that Thanet and his team had engineered about six days earlier.

Once the mission was under way, as each diver came to the opening at the T-junction, he looked right toward the Monk's Series. For once the lack of visibility in the water was reassuring. No clear water from the Monk's Series equaled no new water. For Stanton this was proof that Thanet's diversion scheme had worked. Five or ten minutes apart, each man turned left, heading for the last boys and Coach Ek.

When they reached the chamber, they found two of the SEALs and Dr. Bhak at the waterline. They had observed what the rescuers had been doing in previous days and had readied the boys for departure. Mallinson was first up again. He had expected to take one of the smaller boys, but waiting for him was Coach Ek. No one asked questions, since it was the team's decision, but he and the others found it odd. Everyone had assumed the coach would be last. Harris had given him the option of diving out without sedation, and the coach, through Dr. Bhak's translation, answered that he wanted to go out exactly as the rest of the boys had. Jab right, jab left, and within minutes he was comatose.

Though the coach was a decade older than the boys, several of them were taller than him; once in the water, Mallinson didn't have to strain. He guided him through that longest sump to Chamber Eight, then helped haul him over the rocks with Challen and Rasmussen to Chamber Seven. There, as planned, Jim Warny picked up. Warny, who had been promoted to rescuer, made it back to Chamber Three in record time—the fastest extraction yet.

After handing Coach Ek off to Warny, Mallinson

turned back to Chamber Nine. His final assignment would be to ferry home the very last child, the one they thought might be the smallest. Fitting the boy with dive gear, binding him, and sedating him was a two-person job, and since Mallinson knew the cave far better than Harris, it was decided that he should do the rescue and have Harris as backup.

That morning, Vollanthen had a straightforward dive with the tenth boy, Tee. But just before the home stretch, he was stopped dead in his tracks at the squeeze between Chambers Four and Three. This was the stretch that had given everyone trouble on the first day, and now it was proving problematic all over again. It took him a few minutes to blindly feel his way past the relics of the first rescue attempts—those pumps, hoses, and electrical wires. When he found the narrow opening, he knew there was a wider way through, but it took him a couple of minutes longer to find the part wide enough to fit both his six-foot-two frame and the boy. Wedging himself and Tee through, he swam the boy to Chamber Three. Vollanthen popped out and waited for his mate Stanton.

As is his wont, Stanton methodically picked his way through the passages. There was no hurry, and anyway, hurried divers make mistakes. His abused fifty-seven-year-old body was creaking along. He had Titan in tow, all sixty-six pounds of him. The diver was grateful for the lighter load. Before setting out from Chamber Nine, he had managed to streamline the boy even more by sticking an empty water bottle in the elastic band binding his legs. Now the boy's feet would not scrape the bottom; instead Titan's legs would float—he would be as streamlined as a seal.

The boy was certainly easier to swim with, but instead of stifled breathing, Titan was breathing at a rate of about twenty times a minute. That near hyperventilation began just after Harris had plugged him with the drugs, and Stanton asked if it was all right. Harris said he thought so, and offered the same answer he'd supplied two days prior: "There's nothing we can do about it. As long as he is breathing, just keep going." Stanton was worried the boy would consume all his air. But he soon realized that Titan was only taking little sips of air, and besides, his lungs were so tiny there was no way he'd burn through all that oxygen.

They wound through the maze of passages until finally Stanton neared the nasty squeeze just before the home stretch that had stalled Vollanthen minutes earlier. It was Stanton's sixth time navigating this junction, and each previous time he had approached it differently. Only now he couldn't find the opening. The days' two earlier dives had made visibility go from a few inches to absolute zero.

Stanton had one hand on the guideline and the other on the boy. Minutes went by and Stanton grew frustrated. He was so close to the finish line, he could actually hear the reassuring whir of the pumps thrumming away in Chamber Three.

Titan was tethered, so Stanton let him go and ran his hand over the rough limestone. Finally he found the opening. Relieved, he picked Titan up and swam. As the sound of the pumps grew louder, Stanton was suddenly blinded by what seemed like an airplane searchlight. He kept swimming, bumping the walls toward its source, which turned out to be one of the Chinese rescuers who had jumped in the sump to light up the runway for Stanton. He had instead completely blinded the cranky Brit, and with a good deal of sign language and head shaking was lectured not to do that to the other divers.

It is hard to fault the Chinese rescuer. There were only two more boys left. The dives had gone smoothly, and like clockwork each rescuer had come out about thirty minutes after the other. Chamber Three was stuffed with food and ranking commanders. Rear Admiral Apakorn and a consignment of rice, noodles, and meat—and the less-local fare of Kentucky Fried Chicken and McDonald's—had arrived through the hour-long obstacle course of walking, crawling, and sumping. The admiral was wet and the food was cold, but the mood was jubilant. Mixing in the air with the rank odor of days-old urine was the distinct whiff of glory.

Stanton plopped down next to Vollanthen, wolfed down a cold noodle dish and KFC, and was content. Each of the three divers had made record time. Coach Ek was already completely out of the cave. On the first day it took a full hour for the stretcher bearers and the groups handling the ropes to navigate the boulders, sumps, and chest-deep water between Chamber Three and the cave entrance. On that third day, they hustled Coach Ek through in thirty-seven minutes.

Still, Stanton and Vollanthen were not done with their worries. Both desperately wanted to avoid being in the cave when the rains kicked up. Thanet's dam had held, but with the amount of rain they saw that morning, flooding in the cave was a near certainty. They chatted for about thirty minutes, and right on schedule the rope vibrated and Special Ops Sergeant James Brisbin called out, "Fish on!" alerting them that Chris

Jewell, with the second-to-last boy—Pong—was just 150 or so yards out. The Brits didn't pay much mind and kept talking. Then the rope went still.

On the other end of that line, Chris Jewell's dive had been smooth until suddenly it wasn't. He'd collected the fourth rescuee that day and began chugging toward Chamber Eight. As usual, the child received a couple of additional shots of ketamine on the way. Jewell was 150 yards from Chamber Three when he encountered the same awkward vertical line trap that had stumped Vollanthen and Stanton. But as he attempted to solve this puzzle yet again, he'd gotten into trouble. He was switching Pong from his left hand to his right when he lost the guideline. Visibility was zero in this section— his lights seemed only to illuminate more blackness. It was as if someone had spilled Wite-Out all over his mental map of this section. Jewell had led multiweek cave-diving expeditions into the pit of the Huautla "supercave" in Mexico—a place so deep that it is often referred to as an underground Everest. Yet this sump stopped him in his tracks.

He believes that in the split second he had let go, the tension on the rope sent it pinging up into a snag. Normally it wouldn't be a problem to untangle, but a diver had to find it first—which Jewell couldn't, despite windmilling his arms hoping to snag something. Nor could he find the way forward through that eighteen-inch-wide squeeze. And now, twisting and turning in the water and with no visible landmarks, he had become disoriented. Imagine going into a completely darkened room, spinning around several times, and then being asked to point out the exact direction of an exit the size of a doggie door.

The minutes ticked by; clawing his way on the cave floor with one hand and still gripping the boy with the other, he felt something. It was one of those left-over electrical cables. He assumed he could follow it to Chamber Three; so, hovering low, the boy's air tank serving as a sled, he followed it. But it led him to a chamber he didn't recognize. It felt like the Twilight Zone.

By now more than twenty minutes had gone by. Pong had been submerged in 70-degree water for two hours or so, and while ketamine arrests the body's normal reaction to the cold—shivering and chattering teeth—Jewell could tell by touch that the boy was hypothermic. He was also worried about his air. He only had that one tank. So he hauled him onto dry ground in that mysterious chamber and unfastened his mask and tank. He also took off his own diving gear and began exploring, hoping to spot a landmark. He wondered: *The cave's main passage is more or less a straight shot, so how the hell did I wind up in an unfamiliar chamber?*

Mallinson and Harris made quite the pair, with Harris the better part of a foot taller than Mallinson and thick-set. They were the last two foreign divers back in Chamber Nine. The SEALs had guided the last boy down to them and had begun preparing for their own exits.

As Harris successfully sedated the last boy, Mark, both he and Mallinson realized they had a problem. Mark is thirteen and they expected him to be small, but not *that* small. In fact, he is a hair shorter and even slighter than Titan. (He's the boy in the second navy SEAL video whose face was so thin that it seemed his chin was about to pierce his skin.) He might have weighed seventy pounds with his wet suit on. But the problem wasn't his body, it was his face.

They had one last positive-pressure mask. It was a backup, the older Interspiro model, with a slightly wider build around the face and a flappy skirt—the rubber part that creates a waterproof seal around a diver's face. Harris had already sedated Mark and the men had spent the next twenty minutes yanking and tightening the mask. Mark's nose was squished up against the Plexiglas and there was still a thumb-wide gap under his chin. The mask also had a tricky double skirt. When the men tried to clamp the mask tightly on the boy's face, the skirt would flatten out and water would dribble in. They could not secure a seal, so they decided to dump that mask.

As a secondary backup, Mallinson had brought a pink full mask designed for children—but it was not a positive-pressure mask. Instead of pumping air into the mask—as all the other masks had—this one provided air or oxygen on demand. It would require the boy to be actively breathing, sucking in air. Given the problems encountered by some of the previous boys, Harris and Mallinson were worried that Mark would be the only one to die.

In theory, the safer route would have been suspending the rescue and waiting until the next day for a mask they knew would work. But because of the weather, they didn't have a day. As Mallinson said later, he was "nervous about going in that day anyway, and we knew we didn't have any more time, and we knew this was the last option. If we had not been able to get a seal on that second mask, then potentially we would have had to pull it back and leave him in there for however long."

It proved a tricky problem to solve, even for master improvisers. In fact, they messed with the second mask for an additional thirty minutes to get it snug, by which point so much time had elapsed that the sedation had started to wear off. Harris dosed him again and Mallinson cautiously steered the boy into the water.

The seal was so tenuous that Mallinson half-timed it to Chamber Eight. In the previous dives, the boys had taken dozens of little knocks to the head. The divers had even ripped parts of the SEALs' tank boots—the rubber or foam protective sheaths that slide onto the bottom of scuba tanks—and stuffed them beneath the boys' neoprene hoods to give their heads some cushioning. But Mallinson feared a single bump could dislodge the mask, and the visibility was so poor he wouldn't be able to see it or wouldn't have time to fix it. That first section took him half an hour, nearly twice as long as it had the day before. The boy's regulator was on his right side, so Mallinson gripped the straps on his flotation device with his left hand, curling them right under his chin, so he could hear the reassuring grumble of the bubbles coming up or—at the very least—feel them sliding up his own face. And still the boy's head banged and ground in the tighter spots of the cave. But the seal held.

At Chamber Eight, Mallinson told Challen and Rasmussen not to take off the boy's mask—that seal was too precarious. So, as they switched out his oxygen tank for a fresh one, Mallinson stuck a finger through the quarter-size rubber flap on the side of the mask called a purge valve. It allows a diver to clear water from the mask without taking it off, but doesn't let water in. By sticking a finger in and pushing the rubber flap up, Mallinson could ensure that air did get in. If left unsealed, it would act like an open window on a submarine and cause the mask to start taking in water or free-flowing air, which at the very least would rapidly deplete the tank. At Chamber Six, they did the same routine. But this time the valve stuck, and just as Mallinson put the boy in the water he noticed the open rubber valve and snapped it shut.

After Mallinson disappeared into the water and out of Chamber Nine, Harris put on his gear and followed him as a rear guard. But when Mallinson got to Chamber Four he was stunned to find Jewell waiting there, lonely as a hitchhiker thumbing for a midnight ride. Both divers were a little confused. Jewell had been waiting for Mallinson to come by, knowing he would be the last diver with a boy. But it had taken so long, doubt had begun to colonize more and more of the younger diver's mind. In his exploration of this chamber he had found a used space blanket and wrapped it around the boy, trying anything to keep his temperature from dipping further. A person is considered hypothermic when their core body temperature falls below 95 degrees, and

Pong's would soon be measured at around 84 degrees. Had Pong not been sedated, the hypothermia would have immobilized him and left him unable to respond to stimuli, and could have possibly led to a coma.

The pair realized that Jewell had blindly meandered into an alternative route—shaped like a jug handle—to Chamber Four and had perched himself on the sandy median to look for the telltale lights of a diver. Reoriented, Jewell suited up for the short trip to Chamber Three just as Harris arrived.

"Howzit gaoing?" drawled the big Aussie. Jewell explained how it had gone: not that well. To his credit, Jewell had managed not to panic and had found a way back to a dry area, but Harris could tell he was rattled.

"Do you want me to take the boy?"

The answer was an emphatic *yes*. Harris was already suited up and in the water. Jewell had no intention of being macho about this—the boys were everyone's primary concern. He assembled the tank and regulator, refitted the boy's mask, and handed him to Harris, who began finning to the very spot that had flummoxed Jewell. This time they made it through, heading toward the reassuring hum of the pumps in Chamber Three.

Harris and the boy were pulled out of the water with Jewell right on their tail. Within minutes the Euro-divers followed them out, and cheers went up. There were photo shoots with the divers, the rear admiral, and the SEALs. The Euro-divers were more effusive than the Brits, hugging and backslapping. Erik Brown pulled out a bottle of Jack Daniel's and they swigged from it. Rounds of "Hooyah!" echoed in the chamber. The USAF's Captain Torrel jokingly texted his team that the Euro-divers had stayed with him and were "requesting beer and cigarettes," which caused quite a hoot back at the camp's headquarters.

Stanton wanted a drink, but he wasn't feeling right. Maybe it was the cold food or his infected hands. Maybe it was the prospect of the hour-long return to the mouth of the cave, where the weather was on the outer edge of dangerous.

Almost immediately after the Euro-divers made it to dry land, the teams started packing up. The Chiang Mai rock-climbing team started unbolting the joints they had drilled into the cave walls, the rescuers removed the extra scuba tanks, and the pump workers began to organize equipment so muddy it looked like just another geological feature of the cave floor. All the boys were out. The mission seemed like it was over.

Major Hodges reminded everyone the job was not done, texting the team on its WhatsApp page, "We've got four more SEALs left. Our mission is not complete until everybody's out of there."* They had been working with the SEALs for weeks, and in his mind they were as much a part of this rescue mission as his U.S. unit. So he added, "Hey, good job. Stand by, 'cause we got four more."

Euro-divers Rasmussen, Paasi, and Brown were asked to stay just in case the four SEALs ran into trouble. At the time it seemed like a formality.

Outside, the rains had resumed. The pineapple field was by then a stew of mud and mashed fruit. Journalists—including my producer—were wiping out everywhere, with splats and subsequent groans, followed by futile attempts to wipe mud off gear and bodies. As the mud deepened and seemed to swallow our gear, we had put down toppled pineapple plants to use as mats for surer footing. Despite the leaks coming from Chamber Three, reporters still didn't know if the SEALs were out, or the order in which the boys had been extracted (that secret was held for another

---

* A Wi-Fi router had been set up in Chamber Three, allowing the rescue teams to communicate with headquarters. It apparently also made it possible for some of the workers or rescuers to leak real-time information about the rescue's progress.

six weeks). The parents didn't know, either—which meant that until this moment the parents of the children who had been in the hospital for two days had had no idea if their own children were safe. They were not even allowed to see the first rounds of boys, who were finally conscious and recuperating at the hospital. The government decreed that when all the boys were out, their parents would be allowed to see them at the hospital together.

The Thai government had kept the order of the boys' exit strictly secret, to the extent that when the boys were transferred from the ambulances to the helicopters, squads of umbrella-wielding health workers screened them off (this was also to protect their sensitive eyes from light). Officials hinted that they wanted to avoid an incident in which the euphoria of a family whose child had been rescued could collide with the despair of a family who had lost theirs. It was the kind of policy—benefiting the collective, but temporarily damaging to the individual—that could only be pulled off in a country that lacked some of the transparency of a democracy. Some of the boys were clearly hardy, but until they knew all the boys were out, little Titan's parents had been disconsolate. Their son was the youngest, the least developed, and consequently, they thought, the weakest. They feared that if anyone died, it would be their precious Titan, he of the big smile and big heart.

There was much handshaking on the way out. Stanton and Vollanthen ignored the searing pain of their infected fingers each time another rescuer or worker clamped a hand onto theirs: "It seemed like there were a thousand people lining the last hundred [yards] of the cave. We went from handshake to handshake. I couldn't give everyone a proper shake—had to offer my left hand."

As soon as Stanton and Vollanthen descended the staircase out of the mouth of the cave, cleansed partly of cave junk by the rain, they were asked to visit with the families. It was their first time meeting them. Until that point there had been an invisible barrier between the people who could have lost everything in a rescue and the team who hoped to pull it off. And there they assembled just outside the headquarters, with Josh Morris officiating this unusual and much-delayed union. Most of the thirteen families stood in one line and opposite them stood the lead British divers and their support divers. The Brits had wanted to wait for the Euro-divers, still acting as backup in case something should go wrong with the Thai SEALs, but everyone seemed anxious to get the ceremony under way.

The parents spoke and Morris translated, or tried to. "I had a very hard time, I couldn't keep it together. I was having a hard time getting the words out."

Tears flowed more easily than words. One parent spoke for the rest. But it was Titan's mother who looked directly at Stanton and said, "[I] had died, and now got a second lease on life." Then the Thai parents began a round of hugs—physical displays of affection with strangers are reserved for extreme situations in Thai culture. It was a tender moment that only a member of the tribe of "stiff-upper-lip" Brits could have called "a little awkward"—which Stanton did. Mallinson offered hugs in return, but says he wasn't particularly emotional. And then Titan's mother, petite, with a big-toothed smile like her son, took Stanton aside. She hugged him and—with Morris translating—told him that she had been so worried about Titan; he was so much smaller than anyone and she feared he would struggle. She had been worried the whole time, mad at herself for letting Titan join the soccer team, and now Stanton had brought out her boy alive.

Morris, a life coach as well as a climber, was unable to fully process the moment, his emotions "swim-

388 · MATT GUTMAN

ming around in my head." He began thinking of his own children, close in age to the boys in the cave, and his own antics on ropes and in caves in years past. And after two weeks of keeping it together, he unabashedly wept, right along with the families.

As the rains raged outside, in Chamber Three the Euro-divers, a squad of Thai Navy SEALs, and the U.S. team of pararescuers waited for the SEALs still threading their way back; they were the only rescue personnel beyond the sump at Chamber Three, but there were still dozens of workers packing up equipment in the chambers leading to the cave entrance.

Captain Mitch Torrel, the former hockey player, was in command. Torrel had set his watch. He had assumed that the SEALs would move faster than the divers carrying the boys. During that long waiting period, as some team members munched on the leftover burgers and fried chicken, Torrel took quick stock of the resources available in Chamber Three. He noticed that all the extra air tanks had been removed.

The Euro-divers had been asked by the SEAL commanders to swim in after the SEALs if too much time elapsed. That was a possibly dangerous request: during the entirety of the mission so far, swimming toward incoming divers—either to relay information or check on safety—had been operationally banned. The risk of a collision was too high. But the four SEALs had been immersed in near-total darkness for a week, eating rations that were geared toward sustaining boys, not adult men. They had had no exercise and had been breathing foul, oxygen-poor air. A rescue dive, or in the worst case, a recovery dive for their bodies, was now on the table.

But once again the Thai SEALs' endurance and determination did not disappoint. Following the guideline and the litter trail of the previous rescuers, the SEALs finned toward Chamber Three in a convoy about ten minutes apart. The SEALs had waited two hours after Harris swam off to let the silt settle and to ensure there would be no human logjams or—worse—collisions, at the choke points. They made good time. "Fish on"—the line vibrated and then the first SEAL popped out, followed a few minutes later by the second SEAL. The Euro-divers quietly heaved sighs of relief. Perhaps too soon.

The operation had begun at around 10 A.M., and it was almost thirteen hours later, roughly 10:50 P.M., when one of the three washing-machine-size water pumps bailing thousands of gallons of water an hour from the cave failed. Water started filling the chamber, fast. It was a sequel to the cave's flooding nearly two weeks earlier, when Stanton and Vollanthen had rescued the four hapless workers from Chamber Three. Within minutes it had risen nearly a foot.

At 10:59 P.M. Torrel texted the group on WhatsApp:

Two SEALs out. Pumps broke. Water is filling up fast.

At first his commander, the ever-steady Major Hodges, thought, *Okay, this guy's screwin' with me. He's gotta be jokin'. There's no way that the pumps fail last minute.* But Hodges believed Captain Torrel when the next message chirped up on his phone:

guys are bailing hard.

This meant that rescuers and workers had started to race for the exit, up the forty-five-degree climb to the top of Chamber Three where there was a medical station, and down the mud-slicked slope to the other side of the chamber where it meets the sump to Chamber Two. This was definitely real.

If the water kept rising, the sump between Chambers Three and Two would completely fill again. It was the very spot where Saman Gunan ran out of air and died, and where the Thai Navy SEAL trying to free dive it nearly drowned. And now there were no extra air tanks for the crews not already assigned diving gear. Torrel had estimated that there were about one hundred people in Chamber Three at the time, stripping down and hauling out gear.

The controller for the British divers, Gary Mitchell, texted back seconds later:

On it now guys.

Adrenaline pumping, Mitchell, Hodges, and Anderson conferred—they had come too far to let a mishap like this mar the success of the mission. They could not afford casualties now. Hodges texted Torrel with an order:

pull out before you have to dive out.

Anderson was more emphatic:

get dudes out.

Torrel now shouted, "Landslide! Landslide! Landslide!" to everyone in Chamber Three, making it official. That was the predetermined signal for abandoning ship in case rockfall, air quality, or rising water forced a life-or-death evacuation. It was now official, and dozens of rescuers and rope riggers dropped their valuable gear and began clambering up the slope and "penguin-diving," to get through the sump between Chambers Three and Two before it was sealed with water. Meanwhile, British controller Mitchell jumped on the phone to Thai command, which informed him the power to the pumps had shut down and that they were working on troubleshooting, but needed more time.

As the third SEAL popped up, Torrel grabbed him and hoisted him out of the sump.

Just one more.

The team at headquarters didn't hear back from Torrel immediately. He knew that he had a small pony tank with a mouthpiece that he and his tech sergeant, Ken O'Brien, could use. So, as dozens of rescuers without scuba gear logjammed the sump between Chambers Three and Two, Torrel waited for the last SEAL—nobody was going to be left behind on his watch, even if they had scuba gear.

Anderson and Hodges grew increasingly nervous. At 11:08 P.M., Anderson sent a final urgent text:

GTFO

Get the fuck out. Hodges then called Torrel to speak to him directly, saying, "Hey, you've gotta get out right now. Do not let this turn into a dive mission to exit the cave, because that'll skyrocket the risk level."

He'd barely finished the sentence when Torrel spotted movement in the water.

"Hey, the fourth SEAL just popped out in Chamber Three. We're all headin' out right now."

The USAF Special Tactics team, together with all the SEALs and their scuba equipment, raced for the exit. Torrel and O'Brien scrambled up the forty-five-degree slope to the top of Chamber Three and slid down the mud slide to the sump between that chamber and Chamber Two. The water was now lapping the roof of the sump, and as Torrel bobbed his way through he had to tilt his head back and pucker his lips to sip the little remaining air. Torrel and O'Brien's team, along with that squad of SEALs, waited for the last two SEALs in Chamber Two.

By 11:15 P.M., when an American liaison texted the group that the Thais had gotten the pumps working, the mass evacuation of the entire cave complex was well under way and the water had already sealed the sump between Chambers Two and Three. A long procession of nearly two hundred tired but jubilant souls made its way out of the thin, foul air and into the night.

Inside Tham Luang, there was no one left to save. When it looked like everyone was home free, the Thais stopped pumping. All the equipment left behind—pumps and hoses, tanks and harnesses—would have to be salvaged months later. They would be the only things that had to wait out the monsoon season. Over the next few hours the cave would revert to its natural state for that time of year: an impenetrable river of water that inundated every available air space all the way back to the first chamber.

Chapter Twenty-One
"Life Celebration Party"

As they all proceeded out, the cheers were so intense that Mario Wild of the Chiang Mai climbing team found it "almost cheesy." Rounds of "Hooyah!" drowned out the rain. The rescuers, including Wild, were pelted with shouts of "Heroes!" and "Thank you!" The Austrian had been on rescues before, but "you would never get that kind of recognition and appreciation outside of a place like Thailand."

Josh Morris introduced Wild to the parents. Everyone wept. Vern found Tik out there in the melee of jubilation on that damp night. As he held her, the stiff upper lip of the Brit loosened, and the tears flowed. When Thai Navy SEALs perched on the flatbeds of pickups started heading down the hill past our position in the pineapple field, I splashed through the mud down into the road to savor the moment. Journalists couldn't help but break their own oaths of nonpartisan detachment—we all knew this was too special. We applauded them and howled our own return calls of "Hooyah!"

The courageous Dr. Bhak and his three fellow SEALs emerged to be captured in a snapshot that would soon be beamed around the world. Four wetsuited men in dark sunglasses, thumbs raised skyward, wearing the kind of shit-eating grins you can only get from defying death. It was almost instantly loaded onto the Thai Navy SEAL Facebook page, and from there picked up by media desperate to know the identity of the four brave SEALs. It is doubtful that the rescue would have been successful without their presence in Chamber Nine. Their roles in guiding and calming the boys, managing their food intake, and exchanging the relay of messages with the foreign divers was critical.

Amid the celebrations and back slaps that made Dr. Bhak's head jiggle, an acquaintance asked him if he was going "to the life-celebration party." Life-celebration parties are the Thai version of Irish wakes—celebrations of the dead.

Dr. Bhak asked, "A life-celebration party? Why?"

And his friend solemnly responded, "Ahhh, you don't know yet, there was a life lost in the operation."

None of the divers or the boys had been told about Saman Gunan's drowning death until that moment. It was crushing for Bhak. He learned that Gunan had a family and a career, and was beloved in the SEAL community. In Thailand today, more than that of the boys, the Brits, or any other individual SEAL, Gunan's image is emblazoned everywhere, a testament to the gratitude of a nation for the ultimate sacrifice made for the boys in the cave.

For eighteen consecutive days the cave had endured a human presence. And now, for the first time since June 23, there was not a single human being inside. Last out of the cave were the Americans Torrel and O'Brien. Like the Brits, the Americans would struggle to comprehend their success. That night, wracked with fatigue and disbelief, they headed back to their hotel. The next message on the group text chain that had delivered the hair-raising tick-tock of the final scramble from Chamber Three came the following day at about 5:40 P.M.: "Meeting moved to the pool bar."

Over crisp Laotian lagers and pad thai, the Americans, the Brits, the Australians, and some of the Thai leadership held a rap session about the rescue. Hodges, who had also spent a little time talking with and hugging the families, said later, "It's a surreal-type thing. We don't say zero casualties. Saman Gunan made the ultimate sacrifice for the effort. But wow, this is—this is the stuff of movies."

Anderson, the son of missionaries still working in Ecuador, said the folks around the table agreed that with all those millions of people watching, "we thought, whether you believe in God or not, that something supernatural, something bigger was at work here."

One by one, the soccer players who had been submerged in the monochrome night of the cave blinked open their eyes. Most regained consciousness within a few hours of their rescue. Coach Ek, Titan, Mark, and the others awoke into a world bathed in light. The state-of-the-art ward at Chiang Rai's hospital exploded with it. There were cascades of fluorescent lights, off-white linoleum, white walls and ceilings; even their beds and blankets were white. The only pops of color were the blue numbers at their bed stations and the jade curtains. They had IVs in their arms and surgical masks over their mouths. Two had pneumonia, and all of them were being dosed with prophylactic antibiotics to armor them against microbes attacking their frail immune systems. On average they had lost about five pounds, beefed up somewhat toward the end of their confinement by those calorie-rich MREs. Within hours, most were sitting up, some standing.

The next day their parents, bedecked in the king's yellow, crowded the glass partition that stood between them and their beloved boys. They chirped "We love you," pressing their faces to the glass. Dom's mother's tears smeared the glass. Some couldn't keep the tears of relief and gratitude from flowing. It wouldn't be long before they could actually hold their children's faces in their hands.

Epilogue

The doctors at Chiang Rai General Hospital quarantined the boys for a week. The Moo Pa, in their surgical masks and soft hospital gowns, had the run of the fifty-bed facility. Their parents were permitted to gaze at them through a much-smudged glass partition, but not allowed to touch them. Two of the boys, including Night—who had nearly died on the way out of the cave—were treated for pneumonia; the doctors put all of them on a liquid diet.

The hospital produced a *Brady Bunch*–like video of the kids in a "twelve box" fit onto the screen. Each of the boys sat up in his dazzlingly white hospital bed in the state-of-the-art ward. They thanked the rescuers for their efforts, and the nation for its support. Then the boys talk food: Tee says he wants crispy pork with rice; Adul wants KFC; Titan wants sushi; little Mark asks for steak. They would get none of those things for the time being. Instead they take their meals around a low, square table—sitting on stools in their hospital gowns, bent over hospital food made intentionally bland.

After an agonizing few days, their parents were finally allowed into the ward. Dom's mom buried her face in her son's hair, smelling deeply. She didn't want to embarrass him, but couldn't stop crying. Few of the parents could. The agony of three weeks of uncertainty radiated into frantic fingertips surveying the boys' faces and heads, as if certifying that all the landmarks they remembered were still there—that their eyes weren't deceiving them. Some parents admitted to holding on a little longer than was comfortable for their squirming boys. The boys were now international celebrities, and soon realized that the old rules didn't quite apply. Because after those hugs came the contraband. Titan's mom snuck him some chocolate. Dom complained to his mother that if Titan was the beneficiary of smuggled chocolate he should get some too. His mom figured that if eighteen days in a cave didn't kill him, a little chocolate wouldn't either. So bootleg chocolate appeared under his bedsheets.

It was at the hospital that the boys and coach were

finally informed about Saman Gunan's death. It was gutting news. Of the hundreds of pictures for which they posed, none were more poignant than those of the boys in slippers and hospital gowns crowded around a waist-high, hand-drawn portrait of the fallen rescuer, their faces solemn, hands clasped in front of them.*

On July 17, nine days after the first of the boys had been rushed to the hospital in ambulances, they were all officially discharged. Most had regained the weight they had lost. On that last day each boy, still in his hospital gown, was filmed shuffling through a gauntlet of giggling doctors and nurses. They had been reborn into the world, and in those first ten days hospital staff spoon-fed and coddled the most precious patients in Thailand. When thirteen-year-old Pong went through the line, one of the doctors wasn't satisfied with an athlete's high five and spun the boy around by the shoulders and in for a hug. Everybody laughed. After breaking free, Pong—like all the boys—was asked to say something. For a few seconds he stood there swaying, microphone in hand, unable to get the words out, then wiped his eyes with the back of his hand. Through gasps for breath he thanked the staff "for

---

* The artist's color version of the portrait appears in the photo insert of this book.

always looking after us." Tee, the captain, was more composed, thanking the staff and saying, "I love you very much." Coach Ek also sputtered toward the end of his thank-you, dipping his head and clenching his eyes to dam up the tears. Emotion caused him to abort his speech too. He opted instead for a deep bow, palms pressed together, fingertips reaching high up on his forehead, as a sign of the utmost respect. His was an emotion born of gratitude, but likely also of the trauma of hovering so close to death both while in the cave and during the rescue.

And the following day, more than three weeks after disappearing into the Tham Luang cave, they appeared in public for the first time. Wearing new matching Wild Boars jerseys and shorts emblazoned with a maroon boar racing across the sides, they filed into a Chiang Rai conference hall and took their places on two rows of bleachers. Already seated were Dr. Bhak and the three other Thai SEALs who had stayed in the cave with the boys. An MC offered a long preamble, then asked the boys a series of carefully vetted questions. Coach Ek did most of the talking, explaining in his chirpy voice how they wound up trapped by the water and his decision to pull the boys farther back into the cave for safety. The event was carefully stage-managed, including the boys' public apology to their parents and another tribute to

Saman Gunan. It ended with a deep, collective bow to a portrait of the king. The assembled press were warned that this would be their last chance to see the boys; the MC and doctors implored the insatiable media to give the boys some rest and let them "reconnect with their families in order to heal." But in the coming months it wouldn't be journalists who made demands on the boys, but the Thai government.

The boys finally got to go home. They were now back on their phones, their Instagram pages exploding with new followers. Within weeks, thirteen-year-old Dom would gain hundreds of thousands of followers, though his mother—with her wide, concerned eyes—constantly reminded him that this was all temporary. One of his most "liked" pictures featured a much-belated birthday cake—his birthday had passed unmarked on July 3, the day after Stanton and Vollanthen had found them. As with the other boys who suddenly found social media stardom, Dom's pictures were predominantly selfies and pictures of friends and food.

And then it was time to leave the homes with those freshly made beds with the soccer-ball fleece pillows and the stuffed cats they'd had since early childhood. On July 25, just over a month after they'd entered the cave, the boys traveled just up the road to the gilded

Buddhas and pagodas of the Wat Doi Wao temple. They were to be ordained as "novice monks," while Coach Ek had signed up for a longer stint as an official monk. An apprenticeship as a monk, after all, is one of the greatest merit-making acts in Buddhism. Under the imposing glare of huge dragon gargoyles, a swarm of reporters and Mae Sai officials flocked to the temple. With the temple's abbot chanting ancient rhythms, the boys circumambulated a shrine to the Buddha. They wore matching white shirts and cotton pants; above their heads they all pulled a single bolt of cloth, which formed a saffron-colored ring around the idol. Then it was time for the elaborate head-shaving ceremony. A lock of hair was first ceremoniously snipped and placed in a large banana leaf before clippers did the rest of the work—even shearing off their eyebrows. Tee winced as the monks finished their razor work and washed his bald pate clean.

Shorn heads symbolize a detachment from the daily pageantry of beautifying one's self—a detachment, one could argue, the boys had become wearily accustomed to while stranded in Chamber Nine. The upshot was that they now appeared even younger. Little Titan looked baby-faced and vulnerable as the boys were paraded through the temple grounds in their new uniforms of saffron-colored robes. Soldiers shielded them from the sun with large bamboo umbrellas as photographers snapped hundreds of pictures. This was not the initiation ceremony the average monk receives.

Later, a Mae Sai official credited their survival to meditation, saying that now in their sojourns as monks the boys would be able to improve their meditation practice—which would surely help them later in life. For the next nine days they worked at menial temple jobs—and paid homage to Saman Gunan. It had been more than forty days since they had disappeared into Tham Luang; they had spent only seven of those days at home.

The government stated that it was zealously working to shield them from press interviews. Government-appointed psychologists said they were concerned about the boys' precarious psychological states—recommending a six-month respite from recounting their experiences publicly. But at the same time the junta trotted them out for multiple public events. Even their return to school in early August 2018 was open to the press. A few days later, another event brought journalists flocking back to Mae Sai. After a lifetime of being stateless, Coach Ek, Adul, Mark (the last boy out of the cave) and Tee—the captain—were all granted Thai citizenship.

The showpiece of it all was a massive gala hosted

by the Thai government with a reported guest list topping ten thousand. It was a who's who of the rescuers. The Thai SEALs and Dr. Bhak were there. Vernon Unsworth—still not in possession of a long-term visa—flew in from London. Josh Morris and Mario Wild from Chiang Mai made sure to be there. The U.S. Air Force Special Tactics team was flown in on a special plane from Okinawa. Ben Reymenants and Ruengrit Changkwanyuen and many of the Euro-divers—whose dive shops in the southern resort towns of Phuket and Koh Tao had become enormously popular post-rescue—also turned out. The divers swapped wet suits for business suits. They were all honored—their chests adorned with various pins and medals bestowed on them by Prime Minister Prayut.

At the gala itself, the boys were front and center; their table, cluttered with tubs of ice cream, was next to the prime minister's table. The U.S. Special Tactics team, in dress blues with hair slicked down, came over to take a picture with the boys. They managed a quick explanation of who they were before the boys were bombarded by requests from passing dignitaries hoping for selfies with the boys. To some of the journalists invited it smacked of a depressing carnival show—but the parents noted that it afforded their boys the opportunity to meet diplomats, foreign journalists, and even the prime minster—a big deal for kids from little Mae Sai.

That event also kicked off a number of exhibitions honoring the boys and their rescuers. The Siam Paragon Mall, one of Thailand's largest, featured a mockup of the cave complex replete with a murky, faux-rock tunnel made all the more real by the piped-in sounds of dripping water. With dozens of journalists and onlookers invited to watch, the boys, in matching yellow polo shirts, gamely crawled through the two-foot-high tunnel—using their cell phones as flashlights. It seemed a stunning display of insensitivity on the part of organizers.

In that same exhibition at the Siam Paragon Mall (titled "Tham Luang Incredible Mission: The Global Agenda") there was a replica of Elon Musk's escape pod. Its backdrop was wallpapered with images lifted from the internet of the tech titan in a tux and selected tweets. It was a very public reminder of the rescue's most bizarre episode—one that had not quite ended. Musk couldn't seem to let go of his unsuccessful contribution to the rescue. Divers and diplomats alike quietly let it be known that his sub had been a major distraction. In defending himself, Musk managed to offend Governor Naronsak and Rick Stanton in the space of a single tweet. On July 10 he downplayed Narongsak's importance as a leader and published British diver Rick Stanton's private correspondence with him. In that very same tweet, Musk referred to Stanton as "Dick Stanton." Given the sour note on which their correspondence ended, Stanton assumed it was an intentional insult rather than a typo.

A few days later in an interview with CNN, Vernon Unsworth was asked about Musk's escape-pod concept. In a response that soon went viral, Vern rolled his eyes and a smirk swept across his face. The Brit answered, "He can stick his submarine where it hurts—it just had absolutely no chance of working." Vern seemed to have unwittingly declared war on Twitter maestro Musk, who tweeted this slightly incoherent response: "We will make one of the mini-sub/pod going all the way to Cave 5 no problem. Sorry pedo guy, you really did ask for it." (It's unclear what "Cave 5" is.) Musk never provided any evidence for calling Vern a "pedo." It is hard to quantify how many times that particular tweet was retweeted, because Musk soon deleted it—but it was enough to cause an international uproar in defense of Vern. The firestorm forced a semi-apology from Musk. But then a month later he inexplicably doubled down, reportedly telling a journalist that Vernon Unsworth was indeed a "child rapist" and dared the Brit to sue him. Which is strange, because Vernon's lawyers had already issued preparatory papers alerting the tycoon that a suit was coming.

The four British divers avoided much of the hoopla. Jewell went right back to work. Mallinson took an extended vacation in Spain; I interviewed him in Bilbao, just before he was to (surprise) go cave diving. Vollanthen refused interviews and went back to his IT business. Stanton went right back to retirement, which for him mostly involved traveling, kayaking, and spending time with Amp. Along with Bill Whitehouse of the British Cave Rescue Council, Stanton would start giving a number of lectures. He's not sure about his future in rescue diving, but is content to know that a younger generation of divers—like Jewell—would now have the necessary experience. At the time of this writing, the British divers, the Aussies Dr. Richard Harris and Dr. Craig Challen, Vern, and members of the USAF Special Tactics team still chat on a WhatsApp group. They have become friends, and some of them visit each other once in a while.

During a late-August trip to a soggy Mae Sai, Abbot Ten invited me to meet with him at the Wat Doi Wao Temple. I found the abbot wrapped in bolts of maroon cloth, his legs primly folded beneath him on a plushly cushioned mahogany bench. The soccer team had also assembled there with their parents. Through my translator I gathered that the parents were chafing under the government's demands on their boys' time and energy. As the adults spoke, the boys padded around the temple with the comfort with which most of us move around our bedrooms. Mark, wearing a MAE SAI HEROES T-shirt, posted himself next to Abbot Ten, who wrapped a heavy arm around him as if he was his own offspring. Sitting on the pews or cross-legged on the floor were the parents, the parent's friends, and a psychologist. Catlike, Titan crawled into the lap of a family friend who was hotly debating what the parents should do.

Watching this debate unfold at the temple was strangely reassuring for me. I realized that these boys were cocooned in the safety nets of community: family, temple, school, the Moo Pa team, and—more than ever—each other. I knew that Dom's grandparents and sisters were just down the hill at the amulet shop. Farther down the road, across Highway 1, Nick and Night's extended families were preparing one of their customary dinners for kin—and for friends whom the kids also call their aunts and uncles. Adul had sung in the church choir that Sunday morning, belting out Thai Christian rock along with the other children in attendance. After the weekend, they would all be welcomed back to school, where teachers like Carl Henderson were now perhaps a little more permissive of their quirks—like snacking in class.

During my time in Mae Sai, it became evident that there was a very potent antidote to the deprivations of the cave: adventure. Surviving Tham Luang wasn't enough; these boys insisted on living. Many parents, like Dom's mother, are reluctant to treat them "like an egg in a rock"—that Thai idiom for the suffocating behavior of helicopter parents. "These boys need to be adventurers," she tells me; she reassures herself during Dom's absences that "this kind of thing [being trapped in cave] can happen only once in a lifetime."

Biw got his moped back. Fifteen-year-old Nick got a new midnight blue moped which he uses to zip around town—often with a kid or two riding pillion. Adul, who has become the group's unofficial spokesman, travels all over Thailand without his parents or pastor. They play less soccer now, but spend a lot of time on their bikes exploring the spine of mountains along the Sleeping Princess's profile. None of them has gone back to the cave site.

During my visit to Mae Sai, several of the boys had gone off on an extended bicycle trek around the area. On another night I unexpectedly bumped into the boys at Coach Ek's temple dormitory. They had come to celebrate Titan's twelfth birthday, riding their bikes up the hill to Coach Ek's small apartment and hanging out there with the monk. It was nearly 9 P.M., but as I arrived the boys were already pointing their sleek road bikes back toward home and cheerily bidding Ek goodbye—betraying not a speck of fear. They were, after all, the Moo Pa.

Acknowledgments

When HarperCollins first pitched me to write this book, it was five days after the rescue had ended. In-depth magazine articles had not yet been written, much less any authoritative history or book. The rescue was so unprecedented that there was as yet no frame of reference for it. That necessitated a near-absolute reliance on first-person accounts and the co-operation of interviewees. The first rescuers at Tham Luang on Saturday, June 23—park ranger Petpom and head park ranger Damrong—spoke to me a number of times; when the rescue was over they ushered me through the locked gate in front of the mouth of the cave into Chamber One itself—a space grander than I had ever even imagined. They offered me as much time as I wanted and explained every facet of Tham

Luang. They even introduced me to the park's new pet "Moo Pa"—a "wild boar" that is actually just a dusky pig. Governor Narongsak also sat for an interview and patiently fielded subsequent calls.

I am indebted to Vern Unsworth, who spent dozens of hours coaching me over the phone, through e-mail, and on the apps WhatsApp and Line on the geology, history, and delights of caves in general and on the object of his particular obsession, Tham Laung. He arguably spent more time at the cave site than any other rescuer; he was extraordinarily generous with his time, painstakingly walking me through the events at the cave multiple times.

Rick Stanton, though reluctant at first, subsequently offered me nearly unfettered access, treating me to the quirks of his character and memory—which he called "quite poor." But he kept trying—reminding me to call him about wonderful little tidbits like his "inner tube" and "cockwomble" (see page 92).

The British Cave Rescue Council and its vice-chair, Bill Whitehouse, offered insight into caving in the UK and key access to its sometimes very private divers and staff. I'm also grateful to Chris Jewell and Jason Mallinson, who granted me and ABC News time for interviews—both while they were in England and while on vacation in Spain. Thanks also to the BCRC's Mike

Clayton and to Martin Ellis, for his indispensable maps and Thai caving articles.

The U.S. Air Force offered ABC News (and me in particular) extraordinary access. It opened its doors at Kadena Air Force Base to the network, inviting us to interview officers and NCOs who often spend most of their careers in the shadows. Commanders from Florida to New York to L.A.—including Major Craig Savage—continued to meet with me and answer my unceasing barrage of questions.

Due to the diplomatic and military sensitivities inherent in a project like this, many people spoke to me on the condition of anonymity; my thanks to them. I'm also grateful to Colonel Singhanat Losuya and Captain Padcharapon Sukpang, both formerly of the Thai Thirty-seventh Military District. Captain Sukpang's detailed logbook was most useful for a journalist hungry for exact dates, times, and places.

Thanet Natisri also provided many hours of his time. He had taken copious notes himself, and also furnished me with his satellite maps of the mountain, rain fall totals, and a trove of photos. Thanet smoothly shuttled between the worlds of the Thai and foreign rescuers and offered himself as a capable guide to me as well—introducing me to the many of the people with whom he interacted. Josh Morris, the American owner of Chiang Mai Rock Climbing Adventures, opened up his heart, introducing me to his large team—including Austrian Mario Wild and his brother-in-law, Taw, who played a key role from the second day of the rescue onward.

The Euro-divers were also particularly helpful. Mikko Paasi, Claus Rasmussen, Ivan Karadzic, and Ben Reymenants were the first rescue divers to tell me their stories, which in fact became the genesis of this book. They graciously responded to my calls from vacations and at home in Finland, Malta, Belgium, the Philippines, and Phuket. Ruengrit Changwanyuen, who so expertly navigated the no-man's land between the foreign rescuers and the Thai SEALs, kept answering my calls even as he was at work or dropping his kids off at school. He has a remarkable memory and command of details, names, and events.

Additional thanks go to the American expats living in Thailand, including Bruce Konefe—who gave me an initial briefing on cave diving—the kind and generous Brandon Fox from Mae Sai, and Biw's English teacher Carl Henderson. Thanks also to the U.S. embassy in Thailand, which not only provided information and connections but carefully monitored the safety of its citizens in country.

To the Australians who spoke on and off the record,

thank you. And to Australian anesthesiologist Dr. David Wright, thanks for your time as well.

In 2004, Ben Sherwood, who was then the executive producer of *Good Morning America*, instituted the "Oxygen Rule." More guidance than unbending law, it offered assignment editors carte blanch to deploy reporters, producers, and cameras anywhere in the world where people were at risk of expiring from lack of air. The theory is that there is automatic public interest when humans are trapped in a confined space and at risk of running out of air, with other humans desperately trying to save them. The Oxygen Rule was also a nod to the growing technological arsenal at the disposal of TV news—we could now broadcast from phones, the internet, even social media—enabling reporters to file inexpensively from almost anywhere in the world. After a while the rule became ingrained in ABC News culture. And it partly explains the network's quick deployment and subsequent surge of reporters and staff to the cave site. Sherwood, an author himself, offered early and tremendously useful guidance on quickly getting a project like this started.

Kerry Smith, Senior Vice President, ABC News Editorial, read and commented on every iteration of this writing. Smith possesses twin skills that lend themselves to this process: she is the fastest and most thorough reader I know, and possesses peerless judgment. With her deft touch and decades of experience, she provided thoughtful input throughout. ABC's crack lawyer Maherin Gangat also helped improve the work with each draft she read. I am indebted to ABC News' current president, James Goldston, and his leadership team of Barbara Fedida, Wendy Fisher, and David Herndon for dispatching me on what became a remarkable journey to Thailand, and then generously granting me the time to finish this work. Foreign News Manager Kirit Radiat lived and breathed this story. ABC *20/20*'s Terri Lichtstein brought to life our three, hour-long documentaries about the rescue.

ABC News sent many people to the rescue, all of whom played key roles. Robert Zepeda and Scott Shulman helped get me on air day after day. Correspondent James Longman generously shared reminiscences of his time in Thailand, including one of my favorite episodes in the book—Google Translate's misfired rendering of our driver Nop's effort to inform ABC News that the rescue had started. It resulted in this garbled gem: "never again alpaca." Brandon Baur coordinated ABC News' operation in Thailand; even after ABC pulled up stakes, he managed our network of fixers and translators still on the ground. Led by Than Rassanadanukul, they were our eyes and ears in Thailand, and ultimately became our dear friends. Mancharee Sang-mueang, who juggled journalism with her pharmacy school exams, spent many hours as my language liaison in Thailand, translating documents and articles and ne-gotiating with Thai officials. Carol Isoux accompanied me when I returned to Mae Sai on my last reporting trip there and helped with multiple interviews. Vis-han Chaudhary, with his lightning-fast Google skills, proved a quick and capable research assistant.

The idea for this book was conceived by director of creative development Lisa Sharkey at HarperCollins. The former television news producer reached out to me on Facebook, then assembled a special tactics unit of her own comprised of executive editor Matt Harper and assistant editor Anna Montague. Tremendous thanks must also go to Nyamekye Waliyaya, Andrea Molitor, Mumtaz Mustafa, Leah Carlson-Stanisic, and Maddie Pillari. Sharkey imposed a terrifyingly quick deadline, but made herself available for questions or counsel nearly twenty-four hours a day; she ultimately had a hand in every aspect of the work, from the edit-ing down to the marketing and book placement. Her enthusiasm for this project and her energy kept me going throughout. Sharkey's enthusiasm was matched perhaps only by that of my agents Jay Sures, Albert Lee, and Byrd Leavell—who also served as guides to the uncharted territory of a first book. During the initial month of writing, editor Matt Harper and I likely spoke more frequently and at greater length to one another than to our spouses. Harper threw himself into our shotgun marriage and this project, working nights and weekends even after the birth of his second daughter. It was a dream collaboration with an editor of his intellect and sensitivity.

But my deepest debt is to my wife, Daphna, who was abandoned in the summer of 2018 to go on long-since-planned vacations and parent alone for a couple of months without complaint. For well over a decade after her own career at ABC, she has suffered gamely through the sudden disappearances of her partner to places like Venezuela, the Middle East, Asia, and the wilds of Burbank, California. She was understanding and loving throughout, displaying a generosity of spirit for which I will be eternally grateful. I'm even more grateful to her for our two kids, Ben and Libby—who is old enough to have been a Moo Pa—both of whom keep me laughing and loving every day.

Figures Mentioned in
The Boys in the Cave

THE WILD BOAR SOCCER TEAM

- "Note" Prachak Sutham: Fourteen, tall for his age, and perhaps the most mechanically inclined of the boys. His father works at a local garage.

- "Tern" Natthawut Thakhamsong: Fourteen, defender. Very close to his parents and grandparents.

- "Nick" Phiphat Phothi: Fifteen, cousin and neighbor to Night. Had only joined the team a couple of weeks before they went missing.

- "Night" Peeraphat Somphiangchai: Sixteen, celebrated his birthday the day the boys became trapped in the cave, June 23. Cousin and neighbor to Nick. Joined the Wild Boars only two months earlier.

- "Mick" Phanumas Saengdee: Twelve, goes to the Mae Sai Prasitsart School. Despite being chubby-cheeked, considered one of the better young players.

- "Adul" Adul Samon: Fourteen, originally from Myanmar. He was stateless and lives at the Grace Baptist Church.

- "Biw" Ekkarat Wongsukchan: Fourteen, team goalie. Notorious doodler and snacker in class. He drove the moped from which Coach Ek filmed as the boys rode up to the cave.

- "Dom" Duangphet Phromthep: Thirteen, family owns an amulet shop in Mae Sai, also a 13-and-under captain.

- "Coach Ek" Coach Ek Eakapol Jantawong: Twenty-four, coach of the 13-and-under "Wild Boar" squad. A former Buddhist monk and part of a stateless minority in Thailand.

- "Tee" Phonchai Khamluang: Sixteen (fifteen at the time the boys were trapped). An outgoing team captain.

- "Titan" Titan Chanin Viboonrungruang: Eleven, youngest member of team, had played soccer for four years. Very close to Coach Ek.

- "Pong" Somphong Jaiwong: Thirteen, a natural athlete, had always dreamed of playing for the Thai national team.

- "Mark" Mongkhon Boonpiam: Thirteen, one of the smaller boys on the squad. Was worried about missing exams during the rescue.

BRITISH CAVE RESCUE COUNCIL TEAM

- Rick Stanton: British cave diver who arrived in Thailand with John Vollanthen on June 27. The former firefighter and Vollanthen were the two-man team who located the boys in Chamber Nine. Was ultimately part of the thirteen-diver team that rescued the boys. His girlfriend is a Thai woman from Chiang Rai.

- John Vollanthen: British cave diver who arrived in Thailand on June 27. He's a member of the two-man team, along with Stanton, that located the boys in Chamber Nine. Father of a tween boy and a Boy Scout troop leader, he owns a small IT consultancy. Was ultimately part of the thirteen-diver team that rescued the boys.

- Chris Jewell: World-class British cave diver, works in IT in Sommerset. Was part of the thirteen-diver team that rescued the boys.

- Jason Mallison: British cave diver who brought out five boys. Works as a rope-access technician. Was part of the thirteen-diver team that rescued the boys. Mallinson hauled four of the boys and Coach Ek to safety.

- Jim Warny: Belgian/Irish support diver, positioned in Chamber Six after Karadzic became ill.

- Josh Bratchley: British support diver positioned in Chamber Five alongside Connor Roe.

- Connor Roe: British support diver. Positioned in Chamber Five alongside Bratchley.

- Bill Whitehouse: vice chair of the British Cave Rescue Council.

- Gary Mitchell: Team controller for British divers in Thailand. Also liaised with Thais.

- Martin Ellis: Published surveys and descriptions of Thai caves, including Tham Luang.

- Rob Harper: British veterinarian and frequent caving partner to Vern Unsworth. Harper served as the liaison between the BCRC and Unsworth.

"EURO-DIVERS"

- Claus Rasmussen: Originally from Denmark. Father of three, head of "Euro-diver" team and cave diving instructor at Phuket's Blue Label Diving. Posted in Chamber Eight with Mikko Paasi during the rescue.

- Mikko Paasi: Originally from Finland. Father of two. Diving instructor posted in Chamber Eight during the rescue alongside Rasmussen.

- Ivan Karadzic: Danish, diving instructor now based in Koh Tao, Thailand. Was positioned in Chamber Six with Eric Brown.

- Eric Brown: Canadian dive instructor based in Koh Tao, Thailand. Was positioned in Chamber Six with Karadzic.

- Ben Reymenants: First foreign diver at the cave, was there late on June 26. He put in more guideline than anyone other than Stanton and

Vollanthen. Cave diving instructor at Phuket's Blue Label Diving.

THAI OFFICIALS

- Colonel Singhanat Losuya: Deputy commander of the of Thai Thirty-seventh Military District, Chiang Rai. Worked closely with Thanet Natisri. He played a critical role in compelling Thai leadership to heed warnings of foreign divers' plans.

- Captain Padcharapon Sukpang: Thai Thirty-seventh Military District, Chiang Rai. First military officer in the cave. Saw the boys' names and the handprints on the wall.

- Damrong Hangpakdeeneeyom: Head park ranger at Tham Luang cave.

- Kamon Kunngamkwamdee (Lak): One of the first rescuers on scene, joined Vernon Unsworth.

- Ruengrit Changkwanyuen: GM regional manager based in Bangkok. Cave diver who assisted SEALs during ther first few days and helped coordinate dive efforts until June 3.

UNITED STATES AIR FORCE (USAF)

- Major Charlie Hodges: U.S. Air Force, 353rd Special Operations Group. Leader of the U.S. team at Thai cave.

- Master Sergeant Derek Anderson: 353rd Special Operations Group, Hodges' head of operations. He was largely responsible for drawing up the details of the actual rescue plan.

- Captain Mitch Torrel: 353rd Special Operations Group, commanded USAF operations in Chamber Three, including medics.

- Sergeant Sean Hopper: 353rd Special Operations Group, pararescuer and ropes specialist, posted in Chamber Two.

AUSTRALIAN DOCTORS

- Dr. Richard Harris: Australian anesthesiologist and world-class cave diver. Served as the only doctor on the rescue mission. Arrived July 6.

- Dr. Craig Challan: Australian veterinarian, world-class cave diver. Joined Harris in the cave. Arrived July 6.

SWAZILAND

• Mabuyo Magagula: the tweeter who started Elon Musk's involvement in the cave rescue by tweeting directly at him. Mabuyo had 500 Twitter followers; Musk had 22 million.

PLAYING TO WIN

HOW ALTHEA GIBSON
BROKE BARRIERS AND CHANGED TENNIS FOREVER

KAREN DEANS

ILLUSTRATED BY

ELBRITE BROWN

HOLIDAY HOUSE · NEW YORK

For my Mom and Dad,
whose passions in life have helped inspire my own. I love you.
K. D.

To William N. Brown III (my uncle)
During my childhood you would sit and draw pictures with me
and also bring me art supplies. You couldn't image how much this
meant to me! Thanks for always believing in me.

To Henrietta Smith from ALA
Every time I talk to you, you leave me inspired; you make me feel
like I can fly. Thanks for your encouragement.

To Mr. Frank Stephens—a retired art director from the main branch
of the Free Library of Philadelphia
When I was lost you always helped me find my way. I hope to make
a difference in the lives of young people the way you have made a
difference in my life.
E. B.

Text copyright © 2007 by Karen Deans
Illustrations copyright © 2007 by Elbrite Brown
All Rights Reserved
HOLIDAY HOUSE is registered in the U.S. Patent and Trademark Office
Printed and bound in August 2020 at Toppan Leefung, DongGuan, China.
The typeface is Agenda. The illustrations were done in acrylic paint, pencil, cut and
torn paper, Caran d'Ache crayon, and cardboard attached to 300 lb. watercolor
paper. Some pieces were coated with gesso.
www.holidayhouse.com
1 3 5 7 9 10 8 6 4 2

Designed by Yvette Lenhart

Library of Congress Cataloging-in-Publication Data

Deans, Karen.
Playing to win: the story of Althea Gibson / by Karen Deans ;
Illustrated by Elbrite Brown.—1st ed.
p. cm.
Includes bibliographical references.
ISBN 0-8234-1926-6 (hardcover)
1. Gibson, Althea, 1927—Juvenile literature. 2. Tennis players—United States—
Biography—Juvenile literature. 3. African American women tennis players—
Biography—Juvenile literature. I. Brown, Elbrite II. Title.

GV994.G53D43 2007
796.342'092—dc22
2004052275

ISBN: 978-0-8234-1926-5 (hardcover)
ISBN: 978-0-8234-4853-1 (paperback)

When Althea Gibson was born on August 25, 1927, few would have imagined she was destined for greatness. She was the first child of poor sharecroppers living in South Carolina. Her parents, Annie and Daniel, worked hard, trying to scratch out a living on a cotton farm. Each summer they raised five acres of cotton in exchange for a place to live, which was nothing more than a small wooden cabin. As sharecroppers, they were given part of the cotton they farmed. They sold it to buy just enough food and clothing to get by on.

Times were not good for the Gibson family; so when Althea was three, her parents sent her up North to live with her Aunt Sally, hoping things might improve for her there. Unfortunately, things weren't much better in New York City, so she was sent to live with her Aunt Daisy in Philadelphia. Finally, when Althea was nine, her parents moved to New York and the family settled into an apartment in Harlem. At last Althea was together with her parents, her three little sisters, and her brother.

In the 1930s life in Harlem was tough on kids trying to grow up. Crime was everywhere, and people were poor. Althea was becoming wild. She was either fighting with other kids or skipping school. Sometimes Althea, who loved movies, would play hooky all day in the movie theater. No matter how hard her parents tried, they couldn't keep Althea in line. She had a restless, determined nature that hadn't found a good way to express itself yet.

When she was thirteen, Althea realized she loved to play ball, any kind of ball. Instead of fighting, she started bowling and playing basketball and paddle tennis. She found she was happiest when she was competing in sports, and she was good at just about anything she tried. When she was fourteen years old, a grown-up friend named Buddy Walker recognized how talented she was at paddle tennis and thought she would be good at real tennis. He bought her a used tennis racket, and sure enough, she was a natural!

During the 1940s, tennis was a game played mostly by wealthy white people. The country clubs that had tennis courts would not let black people play. Fortunately for Althea, there was a tennis club for African Americans in New York called the Cosmopolitan Club. It was there that Althea began playing tennis, entering tournaments, and winning matches. But even when she was having success on the tennis court, she was still having trouble at school. She felt that it was a waste of time — she wanted to play tennis instead.

And play tennis she did. There was an African American tennis league called the American Tennis Association (ATA) that held tournaments throughout the year. In 1947, at the age of twenty, Althea won ten straight tournaments in the ATA. She was a champion in the black tennis community, and people were taking notice of her talent. Althea wanted more competition, though. She dreamed of playing in the famous

United States Lawn Tennis Association (USLTA). The only problem was, the officials made it difficult for African Americans to participate. While African Americans were not banned from the league, the championship matches were held at country clubs where they were not allowed. Only one black person, Reginald Weir, had ever entered a USLTA event before. He had lost after the second round.

Luckily there were people who wanted to see Althea succeed. One friend, Dr. Hubert Eaton, invited her to live in his family's home in North Carolina, where she could finish high school, as well as play tennis on his private court. While Althea appreciated his kindness and help, it was difficult for her to live in the South. In North Carolina she was forced to sit in the back of the bus because of her skin color. In those days black people were discriminated against openly in the South, and this made Althea feel terrible.

Though she longed to be in the North again, Althea went to college at Florida A&M, an all-black college where she was offered a scholarship to play tennis. There she began competing in tournaments against white players. In 1950 she qualified to play in the U.S. National Tennis Championships at Forest Hills, the most important USLTA tennis tournament in the United States. Althea was the first African American allowed to play there.

She became a curiosity to many spectators and officials. Some objected to her participation and doubted that she was any good. She gave them something to notice. In her second-round match against a top-ranked player, Althea was headed for victory when suddenly dark clouds covered the sky. Thunder rumbled over the sounds of bouncing tennis balls, and lightning struck down on Forest Hills. Officials stopped the game. When it resumed the next day, Althea lost. But she had broken down a barrier: people were taking her seriously. She was a contender.

During the next few years, Althea struggled. She continued to play in tournaments but did not do well. Some newspapers wrote that she was a big disappointment. Her supporters in the African American community were losing hope that she would ever be number one. Pretty soon she started losing confidence in her game. At one point she was so discouraged, she almost quit playing tennis to join the army. A good friend, Sydney Llewellyn, a taxi driver and tennis coach, convinced her to keep playing. He believed she could be the best in the game, so he coached her and encouraged her to play like a champion.

In 1955 the U.S. government asked Althea to become a goodwill ambassador as part of a traveling tennis team. The team of two men and two women journeyed around the world playing tennis. It was the best thing that could have happened for Althea's career. It allowed her to play lots of tennis while touring Southeast Asia. She saw many sights, met people, had fun, and improved her game.

In 1956, on her way back to the United States, she stopped in Paris, France, to play in the French Championships, the third most important tennis tournament in the world. She won the tournament and became the first African American ever to win a major tennis championship.

But this didn't satisfy Althea. Now she wanted to win the most important tournament of all: Wimbledon. The All-England Tennis Championships at Wimbledon is the oldest, most famous tennis tournament in the history of the game. Althea had played there in 1951 and lost. In 1956 she lost again at Wimbledon and at Forest Hills, but she did not lose hope.

Althea had gone further than any African American tennis player in history, but she didn't believe she had gone far enough. She wanted to be number one in the world. She had the physical skill, but something was holding her back. She continued to play that year, traveling to Australia and Asia for a series of tournaments. She wasn't winning the way she wanted to, however; and on her return to the States, she decided it was time for her to win at Wimbledon and Forest Hills. In 1957 she was going to make it happen.

That year Althea arrived at Wimbledon believing that her time had come to be number one. Although she was nervous, she was confident. All of her hard work finally came together as she won match after match before making it to the final round. On that day she played Darlene Hard, who had beaten her before; but that didn't keep Althea down. She strode onto Centre Court as the queen of England watched from the royal box, and volleyed and smashed her way to the championship. Althea Gibson had won Wimbledon! Later she stood on a red carpet on Centre Court and received a trophy from the queen herself. A few weeks later Althea went on to win at Forest Hills. She was the number one ranked woman tennis player in the world! And if anybody had doubts about her title, she put them to rest when she won the same tournaments again the next year.

Althea Gibson loved to play tennis. She became number one by playing hard and never giving up, even during the tough times. She gladly accepted the help of supportive friends and graciously acknowledged their contributions to her career. Whatever Althea did to break down racial barriers in tennis, she did the only way she knew how: she played tennis like nobody's business.

Author's Note

Not only was Althea Gibson a champion on the tennis court, she was a talented singer and performed twice on *The Ed Sullivan Show*. She also acted in a movie starring John Wayne. In her thirties she took up golf and became the first African American to play for the Ladies Professional Golf Association (LPGA). In her later years she devoted her life to helping children pursue their dreams of playing tennis and golf. She created The Althea Gibson Foundation for this purpose.

Important Dates

1927	Born in Silver, South Carolina, on August 25.
1941	Took tennis lessons at Harlem's Cosmopolitan Club.
1942	Won her first tournament sponsored by the American Tennis Association (ATA), an all-black organization.
1949	Played against white players for the first time.
1950	Competed in the U.S. National Tennis Championships at Forest Hills.
1951	Entered the All-England Tennis Championships at Wimbledon.
1953	Graduated from Florida A&M.
1955–56	Played tennis throughout Southeast Asia as a goodwill ambassador.
1956	Won the French Championships.
1957	Won at Wimbledon and Forest Hills.
1958	Duplicated her wins at Wimbledon and Forest Hills.
1959	Appeared in a film and released a record album.
1964	Began a professional golf career, joining the Ladies Professional Golf Association (LPGA).
1971	Ended golf career; became a professional tennis teacher.
2003	Died in East Orange, New Jersey, on September 28.
2019	A sculpture of Althea Gibson was dedicated on the grounds of the U.S. Open (formerly called Forest Hills).

Selected Bibliography and Further Resources

Althea. Rex Miller. American Masters/PBS, 2014. Film.

Biracree, Tom. *Althea Gibson: Tennis Champion.* Los Angeles: Melrose Square Publishing Company, 1990.

Davidson, Sue. *Changing the Game: The Stories of Tennis Champions Alice Marble & Althea Gibson.* Seattle: Seal Press, 1997.

Gibson, Althea. *I Always Wanted to Be Somebody.* New York: Harper and Brothers, 1958.

Gibson, Althea, with Richard Curtis. *So Much to Live For.* New York: Putnam, 1968.

Gray, Frances Clayton, and Yanick Rick Lamb. *Born to Win: The Authorized Biography of Althea Gibson.* Hoboken, NJ: John Wiley & Sons, 2004.

Learn more about Althea on these websites

https://www.biography.com/athlete/althea-gibson

https://www.tennisfame.com/hall-of-famers/inductees/althea-gibson

https://www.thoughtco.com/althea-gibson-3529145

https://www.wimbledon.com/en_GB/video/index.html